HOW MINING WORKS

2ND EDITION

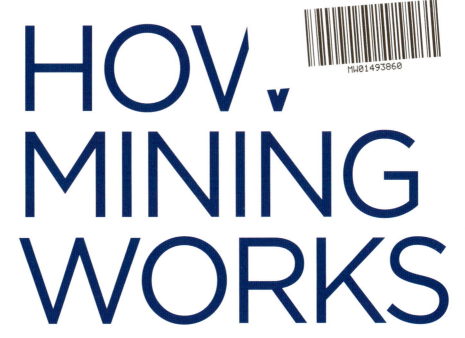

BY W. SCOTT DUNBAR

PUBLISHED BY THE
SOCIETY FOR MINING, METALLURGY & EXPLORATION

Society for Mining, Metallurgy & Exploration (SME)
12999 E. Adam Aircraft Circle
Englewood, Colorado, USA 80112
(303) 948-4200 / (800) 763-3132
www.smenet.org

SME advances the worldwide mining and minerals community through information exchange and professional development. SME is the world's largest association of mining and minerals professionals.

ISBN 978-0-87335-522-3
Ebook 978-0-87335-520-9

Library of Congress Control Number: 2023951878

Contents

Contents

Preface to the Second Edition

After almost 7 years, mining remains interesting. Public discussion about mining has increased in all forms of media, and even in social situations, where one might not expect such a topic to arise. The discussion concerns the effects of mining on a community, on the economy, on geopolitics, and, more recently, the very large demands for metals needed for the renewable energy systems that are part of the energy transition. The connections society has to metals and materials are becoming more visible.

Technical issues do arise in conversation, typically among specialists, and there have been some significant advances in the last 7 years. But it can be said that the focus of these advances is to minimize or mitigate the physical, social, and environmental footprint made by mining in order to supply society with the metals and materials it needs to survive. International bodies, national governments, and society in general are very interested in how such a supply will be provided, what trade-offs will be necessary to make it possible, and whether it is even possible.

Greater understanding of what the mineral resources industry is and how it works was never more important than it is now. Likewise, the industry is changing and needs to understand its relationships, current and potential, with society. One hopes that a closer relationship between society and the metals and materials it uses is developing. Accordingly, other than updating technical descriptions and using better images, the main purpose of this second edition was to completely rewrite Chapter 6, "Mining, Society, and the Environment." Significant updates to Chapter 7 and 8 were also made.

Many of the same people who helped with the first edition also helped with this edition. I have benefited greatly from discussions with colleagues at the University of British Columbia, especially the Norman B. Keevil Institute of Mining Engineering, as well as other departments. People from this global industry provided much information and stories, a few of which are in this book (anonymously, of course). Terese Platten, the lead technical editor for the Society for Mining, Metallurgy & Exploration (SME), did an amazing job of cleaning up jumbled prose and correcting references. The organizational skills of Karen Ehrmann, the copyright permissions editor, are quite something to experience. SME's Book Publishing Manager Melissa Serdinsky kept everything on track and on schedule. This has been a great team to work with. Of course, there can be no unassigned blame and so if there are any mistakes or omissions, it's entirely my fault.

Preface to the First Edition

Mining is interesting. It is a collection of processes that increase the very low concentrations of minerals and metals in the earth to levels that can be used in everyday life. In most cases, the increase in concentration is several orders of magnitude. The fact that it is physically, chemically, and economically feasible to do this is quite amazing. For some metals, the concentration process is like combing through a field of haystacks to find one needle every day for 10 years or more.

What is also interesting about mining is the different kinds of professionals required to make it happen: geologists, almost every kind of engineer or scientist, investment bankers, accountants, economists, health and safety specialists, community relations experts, and others. Furthermore, mining has economic effects at both global and local scales.

The intent of this book is to provide an understanding of the collection of processes, but to do so in a manner that makes the reader want to know more. The only requirement is an interest in science and technology. Very large and very small numbers are used to describe concepts. Some high school chemistry is used to describe processes, but it is not essential for understanding. Appendix A describes all the chemical concepts used.

The book consists of eight chapters. Chapter 1 provides an explanation of how mineral deposits are formed and how they are found. Chapter 2 describes mining methods, the systems and machines used to extract rock containing minerals of interest (called ore) from the earth. Chapter 3 describes a few of the methods used to process the ore and produce metals. Much of mining refers to the production of metals, but nonmetallic minerals are a huge part of the industry. Chapter 4 discusses the particular examples of coal, diamonds, and gravel (aggregates). In addition to minerals and metals, mining and processing ore produce large amounts of waste products that must be managed, often for an indefinite period. The science and technologies applied to this important part of mining are the topics of Chapter 5. Mining also involves people, the communities in which they live, and the government of the country in which the mine is located. As described in Chapter 6, the related issues are important, interesting, and sometimes challenging. Chapter 7 describes the mechanisms for establishing prices of metals and minerals and discusses the legal definition of *ore*. Finally, Chapter 8 presents some ideas about mining in the future.

The majority of the images in the book were generously provided by many companies and individuals. Pictures in a book like this are worth far more than a thousand words, and the attributions under the figures do not begin to express my gratitude and appreciation. Many thanks are extended to those who helped illustrate the many topics. My colleagues Jocelyn Fraser, Mike Hitch, Bern Klein, Rick Lawrence, Allan Moss, and Dirk van Zyl helped with advice and ideas on content, but, of course, any errors are solely my responsibility.

I was encouraged to write this book by Jane Olivier, the books manager at SME, and I am so glad she asked. Diane Serafin edited the manuscript and kept me organized and her head cool despite delays and some software glitches.

The origins of this book lie with Steve Ralbovsky, a former partner at PricewaterhouseCoopers in Phoenix, Arizona, who in 1999 asked me to develop presentations that would explain mining to non-miners attending the PWC Annual School of Mines. For several years Steve twisted the dials and knobs to set the tone and level of the content. The presentations were mostly pictures, but over time I inserted many hidden slides with notes and explanations of the pictures. This was the starting point for this book. Thank you, Steve; it was fun.

In 2007, Simon Houlding of Edumine in Vancouver asked me to make a webcast from the slides, forcing me to learn to speak into a microphone and look into a camera at the same time, pretending that an audience was there. Simon then asked for an online course and quizzes. The information on the hidden slides helped, but good questions that don't frighten the learner are hard to compose. The staff at Edumine also encouraged me to keep the material up to date and to do a Spanish version of the webcast (No puedo. Si, se puede.) Challenges are good. Thank you Simon, Jennifer, Mariana, and Sandra.

Finally, thanks go to my wife Petra, who provided honest feedback, and to my sons and daughters for their comments, ideas, and inspiration.

The Where, What, and How

Have you ever wondered why mines are located where they are?

The ore deposits where mines are situated were formed millions of years ago in particular places by a number of different processes. Most of these processes continue to occur in the earth, but it can take millions of years for an ore deposit to form.

The material in this chapter can help you understand where and how ore deposits form and how geologists go about finding them. It also defines some common terms and concepts used in geology, mining, and mineral processing.

Let's start with what planet Earth looks like. It is the source of all the things we mine.

THE STRUCTURE OF PLANET EARTH

The earth is composed of three concentric shells (Figure 1.1). The crust is the outer shell, 30–50 km thick under the continents and 5–10 km under the oceans. It is underlain by the mantle, which extends from the crust to the interface between the mantle and the earth's core at a depth of 2,900 km. The earth's core, which has a radius of about 3,400 km, consists of an outer liquid core and an inner solid core. The radius of the earth is about 6,371 km.

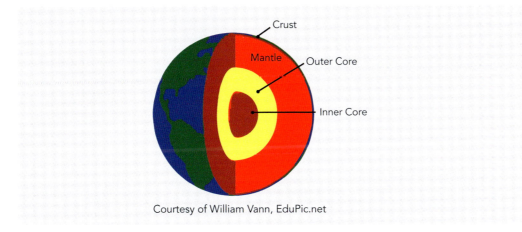

Crust
Mantle
Outer Core
Inner Core

Courtesy of William Vann, EduPic.net

FIGURE 1.1

Structure of the earth

The most interesting aspect of the geological processes involved in forming ore deposits is their scale, both in space and time. The formation of an ore deposit is essentially a concentration process that occurs within the crust and, in some cases, the upper mantle. However, these processes are fueled by heat, generated deep within the earth's core and, in this sense, they are planetary in scale. The result is higher than average concentrations of minerals and metals at depths that are relatively shallow. The earth is a big planet. We have mined only as deep as 3–4 km, barely a pinprick.

Earth History

The earth is very old—about 4.6 billion years old. Table 1.1 shows Earth's history from the time it formed until the present. There are names for particular eons, eras, and periods of Earth's history. The origins of these names is often interesting. For example, the Ordovician period was named after a Celtic tribe called the Ordovices in order to resolve a dispute between geologists who were assigning ages to rock formations in northern Wales as either Cambrian, an earlier period, or Silurian, a later period.

The nickel deposits near Sudbury, Ontario (Canada) are believed to have been formed about 2 billion years ago or 2,000 MYBP (million years before present). Oil deposits in western North America formed between 360 and 400 MYBP, coal

TABLE 1.1

Earth's history

Eon	Era	Period	MYBP*	Major Events
Phanerozoic	Cenozoic	Quaternary	0–1.6	Mastodons and humanoids
		Tertiary	1.6–66	Dinosaur extinction 66 MYBP
	Mesozoic	Cretaceous	66–144	Ore deposits in western North America and South America
		Jurassic	144–208	Coal deposits in British Columbia (Canada) and western United States
		Triassic	208–245	Dinosaurs appeared 230 MYBP
	Paleozoic	Permian	245–286	
		Pennsylvanian	286–320	Coal deposits in eastern United States
		Mississippian	320–360	
		Devonian	360–408	Oil deposits form in western North America
		Silurian	408–438	
		Ordovician	438–505	Trilobites appeared 500 MYBP
		Cambrian	505–543	
Proterozoic			543–2,500	Sudbury (Ontario) nickel deposits
Archean			2,500–4,600	

*MYBP = million years before present.

deposits formed between 150 and 200 MYBP, and several metallic ore deposits in North and South America formed between 60 and 150 MYBP. Dinosaurs appeared about 230 MYBP and their extinction occurred about 66 MYBP, possibly because of the effects of a large meteor striking the earth. The evolution of humans is believed to have begun about 1.6 MYBP.

From Table 1.1, it is evident that geological processes are slow and span hundreds of millions of years. The processes of metal concentration are equally slow and might occur over a few hundred thousand to million years. An equally important concept is that the geological processes that occurred millions of years ago can also occur in the present. Hence the expression:

> The present is the key to the past.

What this means is that the geological processes seen today likely occurred in the past. The climate might have been different, but the earth continued to do what it has always done. Consistency of geological processes over time is a useful concept in mine exploration.

DEFINITIONS

Keep thinking about big spatial scale and very long times as you read through these definitions of terms that are often used in this book.

Chemical Elements

Chemical elements are a pure substance consisting of one type of atom. They are the building blocks of matter. Some examples are given here:

Al – aluminum	Mo – molybdenum
Ag – silver	O – oxygen
Au – gold	Pb – lead
Cu – copper	S – sulfur
Fe – iron	Si – silicon
H – hydrogen	Zn – zinc

Some of the symbols are simply the first letter of the name of the element. Others are derived from the Latin word for the element: **Ar**gentum, **Au**rum, **Cu**prum, **Fe**rrum, **Pl**um**b**um.

Figure 1.2 shows samples of particular elements. Native elements are those found in nature, but these are relatively rare and they are not pure. Elements of economic interest are often combined with other elements.

FIGURE 1.2

Samples of copper, silicon, and gold

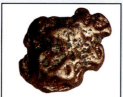

Native copper nugget about 2.5 × 3.5 cm in size. Copper was the first element used by humans. It is combined with tin to form bronze and with zinc to form brass. Copper is the second best conductor of electricity, but it does corrode if exposed to moisture. (The best conductor is silver.)

Silicon. Silicon is one of the most abundant elements in the earth. It combines with oxygen to form quartz, a component of sand that is used to make glass. Since the mid-20th century silicon has been used in electronic devices.

Gold leaf about 1 cm long. Gold is very rare, which makes it valuable. It is also inert and malleable, which makes it useful for jewelry. Gold is the third best conductor of electricity behind silver and copper, but its inertness makes it useful in applications where corrosion would cause a problem.

Photos courtesy of Images-of-Elements.com 2014, CC BY 3.0

Compounds

Elements combine to form *compounds*. A compound contains at least two different elements. Some examples are given in Table 1.2. The numerical subscripts in the chemical formulas denote the number of atoms in the compound. If there is no subscript, then only one atom is present.

TABLE 1.2

Examples of compounds

Compound	Chemical Formula
Salt: sodium (Na) and chlorine (Cl)	NaCl
Silver chloride: silver and chlorine	$AgCl_2$
Vinegar (acetic acid): a bitter combination of carbon (C), oxygen, and hydrogen	CH_3CO_2H
Sulfuric acid: 2 hydrogen, 1 sulfur, and 4 oxygen atoms	H_2SO_4
Copper sulfate: 1 copper, 1 sulfur, and 4 oxygen atoms	$CuSO_4$

Minerals

The definition of a *mineral* is

> A solid, naturally occurring, inorganic compound having a definite chemical composition (of elements) that may vary within limits.

ARE YOU A BIT CONFUSED? WHAT IS A MOLECULE?

A molecule is formed when two or more atoms join together chemically. Compounds form when two or more *different* atoms join together. All compounds are molecules, but not all molecules are compounds. Molecular hydrogen (H_2), molecular oxygen (O_2), and molecular nitrogen (N_2) are not compounds because each is composed of a single element. Water (H_2O), carbon dioxide (CO_2), and methane (CH_4) are compounds because each is made from more than one element.

Some examples of minerals are shown in Figure 1.3.

Rose Quartz, SiO_2. The color, which may vary from pale pink to rose red, is caused by trace amounts of metals such as iron, titanium, or aluminum. The color may be changed by irradiating the crystal with ionizing radiation. (Photo © Anil Öztas)

FIGURE 1.3

Hand-size samples of minerals

Pyrite, FeS_2, is an iron sulfide sometimes known as "fool's gold" because it develops a sheen on its surface that looks like gold. The giveaway is the cubic crystals. Pyrite is the most common sulfide mineral. (Photo by J. Zander 2007)

Chalcopyrite, $CuFeS_2$. Except for the beak and eyes, this carving of an owl is made of chalcopyrite, a common copper sulfide found in most copper mines. The owl is about 6 cm high and is standing on quartz crystals. (Photo by Adrian Pingstone 2002, CC0 1.0)

The following list describes other minerals of interest:

- Hematite (Fe_2O_3), an iron oxide, is the main source of iron. Hematite usually precipitates from standing water collecting in layers. It is typically dark colored, but weathering processes can cause it to become hydrated to form an iron hydroxide, essentially rust, that is responsible for the red color of many soils.
- Sphalerite (ZnS), which is zinc sulfide, and galena (PbS), which is lead sulfide, usually occur together in lead-zinc or lead-zinc-silver deposits.
- Pyrrhotite ($Fe_{1-x}S$, where $0 \leq x \leq 0.2$) is a magnetic iron sulfide with a strange chemical formula that seems to suggest partial atoms. However, this is a

convenient way to show a variable amount of iron. The end member of the series, troilite FeS ($x = 0$), is nonmagnetic.

- The feldspar minerals—$KAlSi_3O_8$, $NaAlSi_3O_8$, and $CaAl_2Si_2O_8$—are silicates that comprise 60% of the earth's crust, a feature that distinguishes the crust from the mantle, which has fewer silicate minerals.

MORE ABOUT MINERALS

The definition of *minerals* earlier in this chapter excludes substances formed by biogeochemical processes. However, there is a close link between the metabolic activities of microorganisms and mineral formation. In fact, microorganisms are capable of forming minerals and particular crystal structures that cannot be formed inorganically. Based on this, H.C.W. Skinner (2005) made the following expanded definition of a mineral:

> An element or compound, amorphous or crystalline, formed through biogeochemical processes

where the prefix *bio* reflects the role of living systems in mineral formation. Skinner views all solids as potential minerals.

Minerals have been traditionally classified according to their chemical and crystal structure. For example, there are silicates and sulfides, cubic and tetragonal crystal shapes, and so forth. This is a static classification system. Robert Hazen and John Ferry (2010) have shown that minerals evolved in a series of three eras beginning with an initial era of planetary accretion more than 4,550 MYBP when there were only about 60 minerals; through a second era of crust and mantle reworking to 2,500 MYBP, after which there were 1,500 minerals; and then into the current era in which living systems catalyze many mineral-forming reactions, resulting in the almost 4,400 mineral species known today. It is a very dynamic system that continues to evolve.

Rock

A rock is a solid assemblage of minerals. A soil is also an assemblage of minerals, but it is not solid. Soils are produced by weathering of rock, a part of the rock cycle that is discussed later. There are many types of rocks, which are also discussed later in context.

Ore Deposit

An *ore deposit* is a solid, naturally occurring mineral concentration usable as mined or from which one or more valuable constituents may be economically recovered.

The implications are that *current technology* and *economic conditions* make economic recovery possible. However, time is an important consideration. Technology can turn an uneconomic deposit into an economic one. However, it can take some time for technology to develop into a practical form. Also, it is long-term economic conditions that govern whether a mine goes into operation and stays in operation. Short-term economic disruptions, such as the crisis in 2008, do not have much effect on mining operations.

This is not a legal definition of ore. The legal definition involves the issue of having the technical and legal ability to extract something of value from it. That legal aspect, and particularly the issue of how an ore deposit is reported to shareholders of a publicly traded mining company, are discussed in Chapter 7.

Base Metal and Precious Metal

A *base metal* is a metal basic to industry and society such as iron, copper, lead, zinc, and aluminum. A *precious metal* is a rare, naturally occurring metallic element of high economic value such as gold, silver, and the platinum group metals: palladium, rhodium, iridium, osmium, and ruthenium.

Ore Grade

Ore grade is the concentration of economic mineral or metal in an ore deposit. A fundamental unit used in defining grade is tonnage. The definition of *one ton* has changed over the years but can be summarized as follows:

- One U.S. ton (1 ton) is 2,000 lb; also called a *short ton*.
- One U.K. or Imperial ton is 2,240 lb; also called a *long ton*—obsolete since about 1965 when the United Kingdom adopted the metric system.
- One metric ton (1 t) is 1,000 kg. The term *tonne* is still used in many countries.

Given that 1 kg is about 2.2 lb, 1,000 kg or 1 t is about 2,200 lb, or about 1.1 times the mass of 1 U.S. ton. (It is close to the mass of the Imperial ton, but there's no need to be confused with an obsolete unit.)

The grade of a base metal deposit is expressed as a mass percentage. For example, a fairly rich copper deposit might have a grade of 0.4%. This means there are 0.004 × 1,000 kg = 4 kg of copper in each metric ton of ore. Alternatively, there are 0.004 × 2,000 = 8 lb of copper in each U.S. ton of ore. This all works out as shown in the following equation—watch the units (remembering that "t" is the metric ton and "ton" refers to the U.S. ton):

$$4 \text{ kg/t} = \frac{4 \text{ kg} \times 2.2 \text{ lb/kg}}{t} = \frac{8.8 \text{ lb}}{1 \text{ t} \times 1.1 \text{ ton/t}} = 8 \text{ lb/ton}$$

The grade of a precious metal deposit is expressed in grams per metric ton or troy ounces per ton, where the common abbreviations are g/t and oz/ton, respectively. The ounces are troy ounces where 1 troy oz = 1.097 standard ounces (avoirdupois) = 31.1 g. For example, the gold grade at the Eskay Creek mine in British Columbia (Canada) was about 0.96 oz/ton or 29.856 g/t, which is a very rich ore. A more typical gold grade is 1–4 g/t for open pit mines, 4–6 g/t for marginal underground mines, and 8–10 g/t for high-quality underground mines.

Because there are 1,000 kg in a metric ton and 1,000 g in a kilogram, there are 1 million g in a metric ton. Thus, if gold grade is given in grams per metric ton, then it is in parts per million (e.g., 6 g/t is 6 ppm).

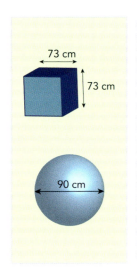

HOW BIG IS 1 METRIC TON OF ROCK?

One cubic meter (m^3) of solid rock weighs between 2.5 and 3 t. This is the density of rock. Assume the density is 2.6 metric tons per cubic meter or 2.6 t/m^3. Invert this to get 1/2.6 = 0.385 m^3/t. This means that 1 t of rock occupies a volume of 0.385 m^3. This volume could be any shape. If it's a cube, then one side of the cube is $0.385^{1/3}$ = 0.73 m or 73 cm. Such a cube is shown to the left.

If the volume is a sphere, then because the volume of a sphere of radius r is given by $4\pi r^3/3$, the radius of the sphere can be computed as $(3 \times 0.385/4\pi)^{1/3}$ = 0.45 m or 45 cm. Therefore its diameter is 90 cm. Such a sphere is shown to the left.

Extend a measuring tape to these dimensions to get an idea of how big these volumes are.

To convert a weight percentage to parts per million, multiply the percentage by 10,000, that is, 10,000 ppm = 1%. Thus 0.4% is 0.4 × 10,000 = 4,000 ppm. Conversion of base metal grades to parts per million is not commonly done, but it serves to show the large difference—three orders of magnitude—between typical gold grades and typical base metal grades.

Diamond mine grades are measured in carats per metric ton (carat/t). Because 1 carat = 0.2 g, a diamond grade of 4 carats/t = 4 × 0.2 = 0.8 g/t or 0.8 ppm.

Rocks and the Rock Cycle

There are many physical and chemical cycles that occur in the earth (e.g., hydrologic cycle, carbon cycle). A rock cycle is illustrated in Figure 1.4.

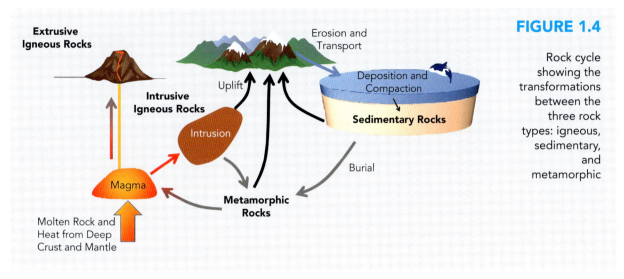

FIGURE 1.4

Rock cycle showing the transformations between the three rock types: igneous, sedimentary, and metamorphic

Three types of rock are shown in the rock cycle: igneous, sedimentary, and metamorphic. Igneous rocks are formed when rock melts at temperatures between 600° and 1,200°C. The melt is called *magma* if it stays within the earth; if it exits the earth, as in a volcano eruption, it is called *lava*. Melted rock is light and flows toward the surface where, if it solidifies within the earth, it is called an *intrusive igneous rock* (or simply an *intrusion*). If it exits the earth, it solidifies very quickly and is called an extrusive igneous rock.

Figure 1.5 shows some examples of igneous rocks. The magma that formed the granite cooled slowly, allowing the large crystals to form. Figure 1.5 also shows rhyolite, which is formed by fast cooling of lava from a volcano, resulting in small crystals. Both rocks have about the same chemical composition. Figure 1.5 includes a sample of basalt, a volcanic rock that contains less than 50% silica and a lot of heavy dark minerals that contain iron and magnesium. While granite is the most common rock in the continental crust, basalt is the most common rock in the oceanic crust. Figure 1.5 also shows basaltic lava that, because of its low silica content, flows very easily. The Columbia plateau in southern Washington state and adjacent Oregon and Idaho (United States) is a large (163,000 km²) "flood basalt" formed during volcanic activity 15–17 million years ago. Hawaii (United States) was formed and continues to be formed by basalt flows.

Sedimentary rocks consist of particles formed by weathering of rock transported by flowing water and deposited either on the earth's surface or at the bottom of a body of water. Unless disrupted by some other process, the deposition results in layers

FIGURE 1.5

Examples of
igneous rocks

Granite

Rhyolite (likely hand-size specimen)
(Courtesy of U.S. Geological Survey)

Basalt (Courtesy of U.S. Geological Survey)

Basaltic lava (Courtesy of U.S. Geological Survey)

of sediments called *strata*. As more sediments are deposited, the layers become compacted and hard. An example of a sedimentary rock formation is shown in Figure 1.6. The sediments forming these rocks were deposited over several million years during the Triassic period, about 200–245 MYBP.

The appearance of the sediments reveals a lot about the environment in which the sediments were deposited. The lower, red-colored rocks in Figure 1.6 are composed of silt (siltstones) and were deposited in a shallow, calm marine environment. The red color is caused by iron minerals (hematite) formed by exposure to the atmosphere, essentially a rusting process. The upper, gray-colored rocks are limestone composed of the skeletal remains of marine organisms. These sediments were deposited in deeper water where the iron present was not exposed to the atmosphere.

Over time, igneous and sedimentary rocks may become buried and subjected to high pressures and temperatures, which cause physical and chemical

FIGURE 1.6

Part of the Virgin Formation, a sequence of sedimentary rocks in southwestern Utah (United States) formed during the Triassic period, about 200–245 MYBP

Photo by M.A. Wilson 2008, CC0 1.0

transformations but do not melt the rock. The result is a *metamorphic rock*. The variety of sources combined with the large possible range of temperatures and pressures gives rise to a wide variety of chemical and crystal textures of metamorphic rocks. Two examples are shown in Figure 1.7.

FIGURE 1.7

Two examples of metamorphic rocks

Quartzite—originally sandstone. (Photo by S. Sepp 2005b; Museum of Geology, University of Tartu)

Gneiss—originally igneous or sedimentary. The bands are caused by high shear forces that cause the rock to behave like plastic. (Photo by S. Sepp 2005a; Museum of Geology, University of Tartu)

Metamorphic rocks comprise about 85% of the earth's crust, and it is estimated that about 10% of the earth's crust is composed of sedimentary rocks. These sedimentary rocks must exist at or near the surface of the earth because at depth, they would be exposed to high pressure and temperatures and become metamorphosed or melt to become igneous rocks.

As may be seen in Figure 1.4, the rock cycle actually consists of a number of subcycles. Rocks are continually transforming from one type to another over millions of years. Although the source of heat is deep within the earth (because of heat generated by radioactive decay), most of the processes shown in the rock cycle occur at depths less than 20 km.

Faults, Fractures, and Folds

The earth's crust is subjected to very large forces that lead to ruptures called *faults* and numerous fractures. There are several well-known faults, such as the San Andreas fault in California where two large plates of the earth's crust slide past one another and cause earthquakes. The Rocky Mountains in Canada were formed by thrusting of sedimentary rocks from the west to the east. Mount Rundle, one mountain in the Canadian Rockies, is shown in Figure 1.8. The uplifted layers can be seen on the left, the east side. The layers have been thrust from west to east along a sub-horizontal fault over existing rocks, a process that has resulted in older sedimentary rocks being placed on top of younger ones, the reverse of what happens when sedimentary rocks are formed.

FIGURE 1.8

View looking south of Mount Rundle (Banff, Alberta), formed by a series of thrust faults

Adapted from Stanley 2006, CC BY-SA 2.0

Fractures ("mini-faults") also result from the large forces imposed on rocks. Fluids with mineral solutions percolate through these fractures and deposit minerals or metals in the fractures, forming veins. An example of small veins containing visible gold is shown in Figure 1.9. (Visible gold is actually a rare occurrence.)

Stresses combined with temperatures and pore water pressure can soften rocks so that they bend easily or actually flow. Layers of sedimentary rocks that were originally horizontal can become *folded*, as shown in Figure 1.10. Folds are easily seen in sedimentary rocks, but they also occur in metamorphic rocks and can form in igneous rocks subjected to stresses in the later stages of magma cooling.

FIGURE 1.9

Gold veins at Brucejack property, northwestern British Columbia (Canada)

Courtesy of Pretium Resources

FIGURE 1.10

Folded sedimentary rocks at Mount Head in the Canadian Rockies

Courtesy of Alberta Geological Survey

CONCENTRATIONS OF METALS

Geochemical models can be used to estimate the average concentrations of metals in the crust of the earth. The details of these models are complex. However, the main observation is that the typical grade of an ore deposit is larger, sometimes much larger, than the average concentration of the metal in the earth's crust.

For example,

> The grade of a typical gold deposit is of the order of thousands of times greater than the average concentration of gold in the earth's crust.
>
> In contrast, the grade of an iron deposit is only 10 times greater than the average concentration of iron in the earth's crust.

These relationships are illustrated for different elements in Figure 1.11. Note the logarithmic scale where one major division denotes an order-of-magnitude difference (e.g., from $10^1 = 10$ to $10^2 = 100$). This scale is used because of the large range of concentrations.

FIGURE 1.11

Comparison of average metal grade (abundance) in the crust with average ore grade

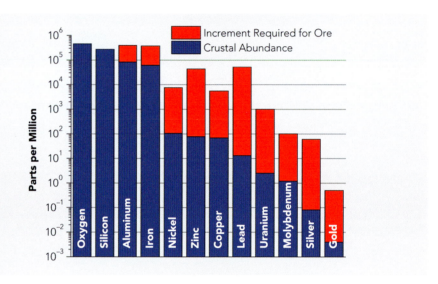

The abundances of oxygen and silicon are shown for comparison:

Silicon ~300,000 ppm, Oxygen ~500,000 ppm

These two elements occur as components of minerals; that is, they are bound chemically to other elements.

The concentration of metal that makes a mineral deposit an ore deposit depends on the current price of the metal because ore is what can be mined under current economic conditions. Thus, as the price increases, less metal concentration is required to make an ore deposit. However, relative to the concentrations in the earth's crust, the changes are not significant.

The main point is that something has to happen in the earth's crust to make a difference in the concentration of a particular metal, that is, to make an ore deposit. The next sections describe some of the processes that make ore deposits.

ORE-FORMING PROCESSES

An ore-forming process is one that concentrates metals or economic minerals above the average crustal concentrations shown in the previous section. Ore deposits can form at any point in the rock cycle, leading to the very broad classification of ore deposits into igneous, sedimentary, and metamorphic deposits. However, igneous processes are the most prolific in terms of the variety and value of mineral deposits formed. One sign of this lies in the chemistry of metallic minerals shown here:

Pyrite: FeS_2 Chalcopyrite: $CuFeS_2$

Sphalerite: ZnS Chalcocite: Cu_2S

Molybdenite: MoS_2 Millerite: NiS

Galena: PbS Pentlandite: $(Fe,Ni)_9S_8$

Argentite: Ag_2S Linnaeite: Co_3S_4

The common element is sulfur (S), which is produced by volcanoes. The locations of current volcanoes are shown in Figure 1.12.

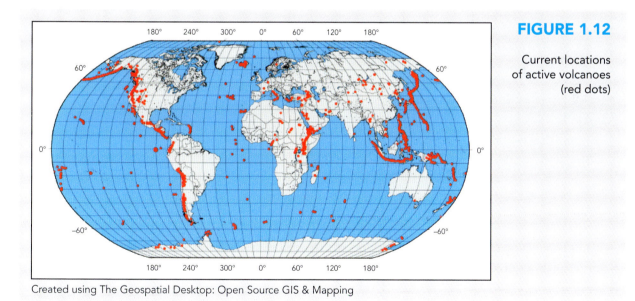

FIGURE 1.12

Current locations of active volcanoes (red dots)

Created using The Geospatial Desktop: Open Source GIS & Mapping

It is interesting to compare the map in Figure 1.12 with a map showing the locations of "giant" ore deposits in Figure 1.13. The correlation with volcanic activity is clearly evident, especially on the west coast of North and South America.

FIGURE 1.13

Locations of "giant" ore deposits. The purple lines are zones where significant generation of metallic minerals occurs.

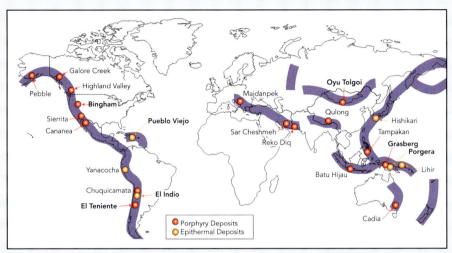

Source: Richards 2013, © Springer Nature

Igneous–Hydrothermal Ore Deposits

If ore deposits are spatially associated with volcanoes, they must be a result of magma generated at depth under a volcano. Figure 1.14 shows the processes that occur when magma forms. Note the size and scale. It is a very "busy" drawing because many things happen when rock melts; the liquid magma emits hot gases and fluids, leaches chemical elements (including metals) from the solid rock, and exerts pressure against solid rock causing fractures and other disturbances. (Molten rock can really mess up a region.)

FIGURE 1.14

Ore-forming processes that might occur where magma forms

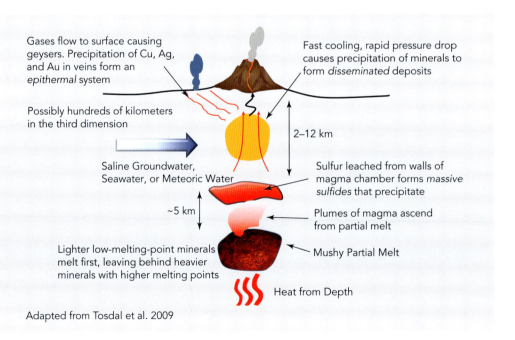

Adapted from Tosdal et al. 2009

At such large depths, the rock mass is under tremendous pressure, which raises its melting point. The minerals in the rock have different melting points, and a *partial melt* forms as the temperature increases. Feldspars and quartz have the lowest melting points (about 600°C) and are the first to become liquid, resulting in a felsic magma that ascends from the partial melt. Left behind are crystals of the iron-rich minerals olivine and pyroxene that have a higher melting point, about 1,000°C. A basaltic magma forms once that temperature is reached. Thus, the less dense felsic minerals, which could contain metals of economic interest, are separated from the heavier minerals in a rock mass.

Magma may melt enough rock to form a chamber where it accumulates. Sulfur leaches from the walls of the chamber and forms sulfides that precipitate once the solution becomes saturated, resulting in a massive sulfide deposit. Nickel deposits at Voisey's Bay (Newfoundland and Labrador, Canada) and Norilsk (Russia) formed this way.

The addition of saline groundwater, seawater, or meteoric water (rainwater or snowmelt) to magma creates a hydrothermal system, a hot (>200°C) brine that can dissolve and transport large quantities of metals and other elements. As a hydro-thermal solution moves upward, it cools and undergoes a pressure decrease causing the dissolved metals to precipitate as minerals at temperatures between 50° and 500°C. Disseminated mineral deposits form at higher temperatures, and epithermal deposits form in veins at lower temperatures. (Epithermal deposits are a source of gold.) A volcano may erupt and/or a geyser may form if the hot fluids reach the surface. The geysers at Yellowstone National Park (Wyoming, United States) are an example of an underlying epithermal system.

Porphyry copper deposits are disseminated deposits formed when metallic minerals are deposited from hydrothermal solutions within hairline fractures and larger cracks. Porphyries are low-grade (0.5%–2% Cu), high-volume deposits with smaller amounts of other metals such as molybdenum, silver, and gold. They are the most common source of copper.

PARTIAL MELT IN THE KITCHEN

One kitchen example of a partial melt is frozen juice. The sugary juice has the lowest melting point, between –30° and –40°C, and thaws first as the temperature increases. The water in the juice has the highest melting point (0°C) and is the last to thaw. This is also why ice cubes made of frozen juice taste so good at first—the sugary juice is the first to thaw—and then taste exactly like frozen water.

Given a partial melt, not all of the processes shown in Figure 1.14 would occur. For example, a magma chamber may not form due to poor circulation of magma from the melt, or a hydrothermal/epithermal system may not form due to lack of water. Also, to generate sufficient mineralization to form an ore body, the supply of molten rock must be continuous over a long time period, a few hundred thousand or a few million years. There may also be several phases of more intense magmatic activity.

Ore Deposits and Plate Tectonics

If ore deposits are associated with the formation of magma or molten rock, where does rock become hot enough to melt? As previously stated, rock completely melts at temperatures of about 1,000°C. On average, the temperature in the earth increases at a rate of about 20°C per kilometer such that melting should occur at a depth of 50 km. However, rock in the earth is known to be solid at depths much greater than 50 km. The reason it does not melt is that pressure also increases with depth and increased pressure inhibits melting, resulting in an increase in the melting temperature. Thus, to form magma, something must overcome this increase in melting temperature. This happens at plate boundaries.

The crust of the earth is divided into a number of thick plates, as shown in Figure 1.15. The plates move relative to each other at different types of boundaries, shown as thicker black lines on the map. The plates move as a result of flow in a deep region within the earth's mantle called the *asthenosphere* where temperatures are such that the rock is viscous. The overlying plates are rigid and include the crust and upper part of the mantle, a region known as the *lithosphere*. Under oceans, the lithosphere is about 50–100 km thick while the thickness of the continental lithosphere ranges from about 40 to as much as 200 km. (The upper 30–50 km of continental lithosphere is continental crust.) The only distinction between the lithosphere and asthenosphere is that of rock behavior. In response to stress, the lithosphere deforms elastically (recovers its original shape), whereas the asthenosphere undergoes viscous flow.

Figure 1.15 shows several types of boundaries between the plates. Mid-ocean ridges are large mountain ranges in the ocean, and subduction zones are where a heavier (iron-rich basalt) oceanic plate goes under a lighter (silica-rich granite) continental plate. Island arcs are also subduction zones, but the overriding continental crust happens to be below sea level. Continental rift zones are on-land versions of mid-ocean ridges. The Basin and Range Province of Nevada and Utah (United States) is an old (17 million years) continental rift, whereas the East African rift zone, and the Gulf of California are current rift zones.

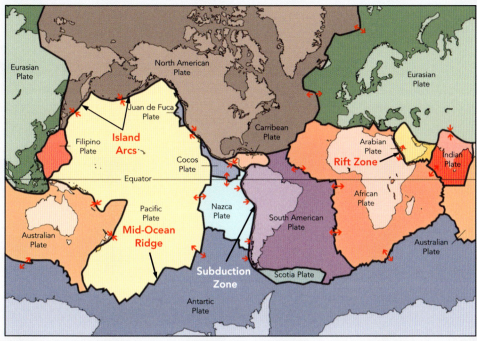

Courtesy of Earthquake Science Center, U.S. Geological Survey

FIGURE 1.15

Plates of the earth's crust. The red arrows show the direction of relative movement of the plates.

The study of the movement of these crustal plates is called *plate tectonics*, or "plate crashing" in polite company.

Mid-Ocean Ridges

The mid-ocean ridge system is an undersea mountain range, the longest continuous feature on earth (80,000 km long and it looks like the seams on a baseball). The boundary between the two plates is a large fissure in the ocean floor that causes a reduction of pressure on the underlying rock, a basalt. Combined with the introduction of seawater, the melting point of the rock is lowered and, as a result, magma forms, flows up through the fissure onto the ocean floor, and solidifies to form new oceanic crust, part of an oceanic plate (Figure 1.16).

HOW LONG DOES IT TAKE TO FORM AN ORE DEPOSIT?

Geochronological (radioactive age-dating) and geological data suggest that a disseminated or epithermal deposit can form in a few thousand to a few hundred thousand years (a "geological instant"). However, these types of deposits are associated with an igneous–hydrothermal system that is formed and sustained by repeated intrusive events over a much longer time, perhaps as much as a few million years.

FIGURE 1.16

Cross section of a
mid-ocean ridge

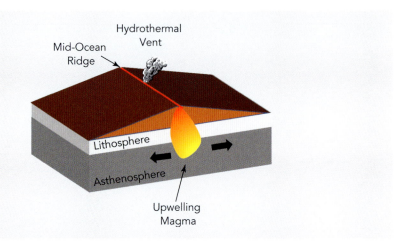

As shown by the thick arrows in Figure 1.16, the oceanic plate moves away from
the ridge along the asthenosphere. The rate of movement is dependent on the rate
of magma generation; for the mid-Atlantic ridge, it is 25 mm/y, while in the Pacific
Ocean, it varies between 80 and 120 mm/y. (Human fingernails grow at an average
rate of 36 mm/y.)

Cold seawater circulates through the many cracks surrounding the fissure and
comes into contact with hot rock or magma at depth. Chemical reactions between
the seawater and hot basalt transform the cold seawater from an alkali solution
containing dissolved oxygen, calcium, magnesium, and sodium salts to a hot, acidic
solution containing dissolved metals, hydrogen sulfide gas, carbon dioxide, and no
oxygen. The basalt is altered to clay-like minerals. The hot acidic seawater dissolves
metals such as copper, zinc, and iron out of the hot rock, rises to the seafloor, and
emerges as a jet flow out into the ocean. The result is a field of hydrothermal vents
(Figure 1.17).

When the hot fluid jet mixes with cold ocean water, the dissolved metals combine
with the hydrogen sulfide to form a suspension of fine-grained sulfide minerals.
The minerals precipitate to form chimney-like structures, sometimes several meters
tall, from which the jet emerges. In very hot vents, the suspension of fine-grained
minerals is composed mainly of iron sulfides that give the jet a black color and the
vent is called a *black smoker*. A white smoker is caused by lighter-colored minerals
containing calcium and silicon that precipitate at lower temperatures.

The chimneys eventually collapse to form a mound, or the minerals may accumulate within a local depression, resulting in a volcanic massive sulfide (VMS) deposit.
Some examples of VMS deposits are shown in Table 1.3. Note the high grades of
silver and gold—these are very rich mines. Historically, VMS deposits account for

FIGURE 1.17

Sully vent in the Main Endeavour Vent Field, northeast Pacific Ocean. Tubeworms cover the base of the vent.

Courtesy of National Oceanic and Atmospheric Administration, PMEL Earth-Ocean Interactions Program

Mine	Reserves, Mt	Cu, %	Zn, %	Pb, %	Ag, g/t	Au, g/t
Neves-Corvo, Portugal	270	1.6	1.4	0.3	30	—
Rio Tinto, Spain	250	1.0	2.0	1.0	30	0.22
Kidd Creek, Ontario (Canada)	149.3	2.89	6.36	0.26	92	0.05
Mount Lyell, Tasmania	106.8	1.19	0.04	0.01	7	0.41
Crandon, Wisconsin (USA)	61	1.1	5.6	0.5	37	1.0
United Verde, Arizona (USA)	30	4.8	0.2	—	50	1.37
Eskay Creek, British Columbia	4	0.33	5.4	2.2	998	26.4

TABLE 1.3

Volcanic massive sulfide deposits

Source: Galley et al. 2007

27% of Canada's copper production, 49% of its zinc, 20% of its lead, 40% of its silver, and 3% of its gold. VMS deposits are also significant sources of cobalt, manganese, cadmium, indium, gallium, and germanium.

All of the deposits in Table 1.3 are in areas where there is currently no mid-ocean ridge. An important piece of information in the exploration for VMS deposits is knowledge of where mid-ocean ridges existed in the geologic past.

The circulation of water from the cold ocean to the hot rocks at depth and back to the cold ocean provides a mechanism for metal enrichment of the oceanic plate. Given that the plate moves away from the ridge, one might ask whether this

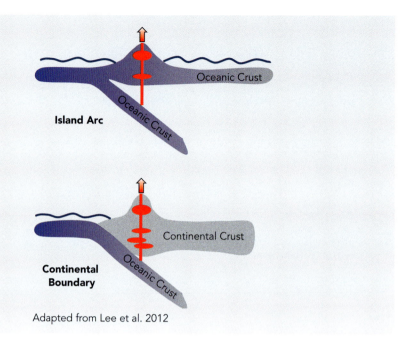

FIGURE 1.18

Two types of
subduction zone,
an island arc and
a continental
boundary

Adapted from Lee et al. 2012

enrichment is lost or if there are other opportunities for this enrichment to be concentrated into ore bodies. Indeed, there are.

Subduction Zones

An oceanic plate moves along the asthenosphere much like a conveyor belt. The plate is composed of dense iron-rich basalt so that at a subduction zone it goes under the lighter silica-rich granitic continental plate. Two types of subduction zone are shown in Figure 1.18. An island arc is where the overriding continental crust is below sea level, and a continental boundary is where the overriding continental crust is above sea level.

At a depth of about 100 km, dehydration of water-bearing minerals in the oceanic plate decreases the melting point of the rock, leading to magma formation. Fluids rising from the subducted plate dissolve the metals in the metal-enriched oceanic plate and the metals in the overlying continental plate. The igneous–hydrothermal processes illustrated in Figure 1.14 then concentrate the metals to form ore deposits.

Plate movement has been going on for millions of years, and it is possible to "run plate movement in reverse" to determine the configuration of the continents up to about 600 million years in the past. The geography of western North America 75 million years ago is shown in Figure 1.19. The entire region was an island arc subduction zone (arrows show direction) with volcanic activity all along the coast that formed the many copper-gold ore deposits in western North America.

FIGURE 1.19

Western North
America 75 MYBP

Kula
Plate

Farallon
Plate

Late Cretaceous
(Santonian) --
85 Ma (87-83)

0   600 Mi
0   1000 Km

Image © 2023 Colorado Plateau Geosystems Inc.

Sedimentary Ore Deposits

Sedimentary ore deposits can form in one of two ways. Dense minerals, such as gold, may be eroded from a "mother lode" in the upstream area of a river. They are then transported to the river and deposited in the sands and gravels of the riverbed, forming an *alluvial* placer deposit. Excavation and processing of these deposits is done to extract the gold using methods that depend on the large density difference between gold and other minerals, gold being relatively heavy. A placer gold mining operation in the Yukon (Canada) is shown in Figure 1.20.

A second mechanism for the formation of a sedimentary ore deposit is where the mineral precipitates from the ocean during the formation of sedimentary rocks. Initially, these sediments are saturated with water, but when buried by more sediments or when the water body evaporates, processes that take place over millions of years, they lose water.

An evaporite deposit is formed when the ocean evaporates or recedes. Salt deposits in the southern United States are one example of such deposits. Another

FIGURE 1.20

Placer mining in Scroggie Creek, Yukon, Canada

Courtesy of Pacific Ridge Exploration Ltd.

example is the potash deposits in central Canada, which were covered by an ocean between 300 and 400 million years ago. Sylvite (potassium chloride), the main mineral in potash, precipitated from the ocean, forming very thick layers of potash. Subsequently, sediments were deposited to form layers 1 km thick over the potash layer.

Some of the most interesting mineral crystal structures form in sedimentary environments.

Metamorphic Deposits

Metamorphism occurs when temperatures are between 200° and 800°C and pressures are higher than 300 MPa, or megapascals. (Car tire pressure is about 50 psi or 0.35 MPa.)

There are two types of metamorphism:

1. Contact metamorphism where the heat from a magmatic intrusion deforms the rock and may alter its chemistry
2. Hydrothermal metamorphism where hot fluids emitted from magma leach metals from fractures in rock and transport them to a deposition zone

In neither case does the rock melt. The distances over which these types of metamorphism occur are large, perhaps 10–20 km. Examples of mineral deposits formed by metamorphism are given in Table 1.4.

Asbestos	Formed by hydration of ultrabasic (<45% silica) igneous rock	
Gems	Jade, garnet, emeralds, beryls formed by high-pressure contact metamorphism of igneous or sedimentary rocks	
Talc and soapstone	Formed by hydrothermal metamorphism of limestone (calcium carbonate)	
Barite	Formed by hydrothermal fluids leaching barium from silicates of sedimentary rocks (hydrothermal metamorphism)	
Uranium	Formed by hydrothermal fluids leaching uranium from underlying rocks in the Athabasca Basin, Saskatchewan (Canada)	

TABLE 1.4

Mineral deposits formed by metamorphism

Oxide Deposits

Once an ore deposit forms, rock layers overlying the deposit may be removed by erosion. This exposes the deposit to air and water, which leads to changes in the minerals of the deposit.

Figure 1.21 shows the oxidation process for a porphyry copper deposit. Primary sulfides in the hypogene or deep zone, such as chalcopyrite, break down into secondary copper sulfides and carbonates to form a supergene zone. Above the water table, oxygenated water, derived from rainwater, trickles through fractures in the rock and dissolves or oxidizes the minerals in the supergene zone to form a layer of copper oxides. Accumulations of secondary sulfide and oxide minerals are typically much lower grade than primary sulfide deposits. However, the oxide layer above most porphyry copper deposits is a significant source of copper (see Chapter 3).

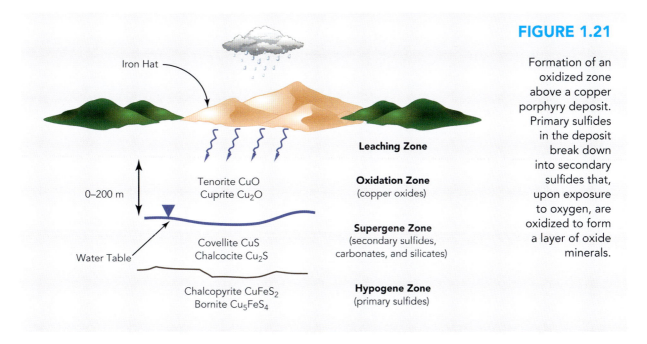

FIGURE 1.21

Formation of an oxidized zone above a copper porphyry deposit. Primary sulfides in the deposit break down into secondary sulfides that, upon exposure to oxygen, are oxidized to form a layer of oxide minerals.

Limonite, an iron hydroxide mineral that is a product of the oxidation process, may deposit above the oxidation zone. Such a deposit is known as a *gossan* or "iron hat." It is typically rust-red in color and is used by prospectors as an indicator of underlying mineralization.

Deep weathering caused by high rainfall in tropical climates can leach metals from underlying rocks to form a surface soil layer, called a *laterite*, which may contain relatively high concentrations of metals to form a residual deposit. Laterites formed from rocks that have a low silica content are sources of nickel, cobalt, and manganese, whereas laterites formed from high-silica rocks contain bauxite, an aluminum mineral.

MINE EXPLORATION

Several methods are used to find ore deposits. An important prerequisite for finding them is an understanding of the manner in which an ore deposit is formed. Several methods are used, in increasing order of cost per square kilometer:

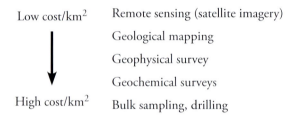

It is important to understand the role of exploration in mining:

- Every mining company must find more tons of ore to replace each ton of ore it extracts. Otherwise, ultimately, it will cease to be in operation.
- The alternative to exploration is acquisition of mining companies with mineral rights to known quantities of ore. Acquisition is often more expensive than exploration for the same amount of ore because there is much less uncertainty about the known quantity and value of the ore. Put another way, someone else has taken on the risk of exploration failure and wants to be rewarded for that.

Mining exploration programs can fail to find any ore of value. The failure rate is high, likely greater than that of research and development in a pharmaceutical or advanced technology company. However, the need to do exploration in the face of such high risks is entirely analogous to the need for pharmaceutical or advanced technology companies to perform high-risk research for developing products to stay relevant in a competitive market.

There is considerable intellectual challenge in exploration—the problem is to find out if and how geological processes millions of years ago led to a mineral deposit. There is a "joy of the hunt" aspect to exploration. In addition, exploration costs are tax deductible in most jurisdictions, and the possible financial rewards are

AN EXAMPLE OF GEOLOGICAL "BIG PICTURE" THINKING

Mine exploration is usually about what is on or in the ground. However, sometimes it can be about what is *not* there. The map below shows copper-gold occurrences in central British Columbia, Canada. There is a significant gap in these occurrences between the Mount Polley and Mount Milligan deposits. Given the broad scale over which ore-forming processes occur, this is unexpected. It suggests that something is covering up copper-gold occurrences in this part of the province. Indeed, the region is covered by a layer of glacial sediments. The area is an exploration target and the composition of the glacial sediments is of interest because they may contain minerals scraped off the underlying rocks by movement of glaciers.

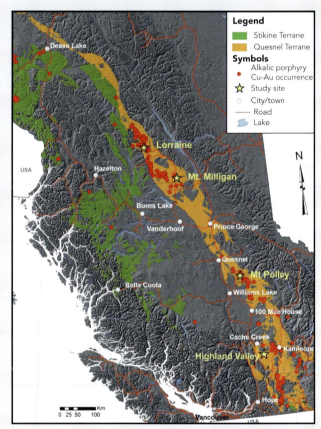

Copper-gold occurrences and mines in central British Columbia, Canada
Source: Bouzari et al. 2010

considerable. These intellectual, primal, and financial incentives are the reasons a junior exploration company engages in the high-risk business of exploration, typically using money obtained by selling shares in the company. Once a significant amount of ore has been proven on a property, the company can sell the mineral rights to a larger mining company with the skills and resources necessary to mine and process the ore. The proceeds from that sale can then be used to fund the next exploration project. This can be a rewarding career, but it is not for the faint of heart.

Figure 1.22 shows a satellite image of the Great Slave Lake in the Northwest Territories of Canada. The colors are pseudo-colors derived from a grayscale image by mapping each pixel of the image to a different color depending on its gray intensity. What interests geologists is changes in the color that might indicate the existence of geological features (e.g., faults and fractures) conducive to the concentration of minerals.

FIGURE 1.22

Satellite image of Great Slave Lake in the Northwest Territories of Canada

Courtesy of National Aeronautics and Space Administration

Geological Mapping

Once a favorable area has been identified, geologists go into the field and make a map of bedrock exposures, or outcrops, in the area to discern any spatial relationships that indicate the presence of economic deposits of minerals. Representative geological mapping requires ensuring that a rock exposure is bedrock, not a "bed boulder." (Usually, the map is made with the aid of portable global positioning devices.) An indicator might be the actual mineral itself, but more likely it would be minerals that are associated with magmatic or hydrothermal activity that could have formed a high concentration of economic minerals. Structural controls on the location of mineralization, such as faults or folds, would also be located on a geologic map. The manner in which the minerals in the rock are affected by the forces that led to the formation of faults or folds may also be useful for determining details on the distribution of mineralization.

Geophysical Surveys

Geophysical surveys might be performed to find changes in physical properties in the area. Examples of physical properties might be the earth's magnetic field, the density of rocks, or the response of the rocks to a radio signal. The presence of metals in the rock would alter the earth's magnetic field intensity and would also change the response to a radio signal. The force of gravity over light rocks would be less than the force of gravity over surrounding denser rocks. All of these properties can be measured with sophisticated and sensitive instruments. The result is a map of changes in these properties and it is the significant changes, called *anomalies*, that might indicate something of interest is present.

Figure 1.23 shows an airplane with a "stinger" containing geophysical instruments attached to its tail.

FIGURE 1.23

An airplane with a "stinger" containing geophysical instruments

Courtesy of Sander Geophysics

Geophysical data can also be collected on the ground. Figure 1.24 shows a map of magnetic data collected on the ground along the black grid lines shown. Red denotes high values of magnetic intensity, green represents intermediate values, and blue shows low values. In this case, the high values of magnetic intensity were caused by the presence of the magnetic iron sulfide pyrrhotite, which is known to be associated with gold. Thus, the areas with high magnetic intensity are potential targets for drilling.

FIGURE 1.24

A magnetic anomaly map resulting from data collected on the ground along the black grid lines shown (red denotes high values of magnetic intensity; blue denotes low values)

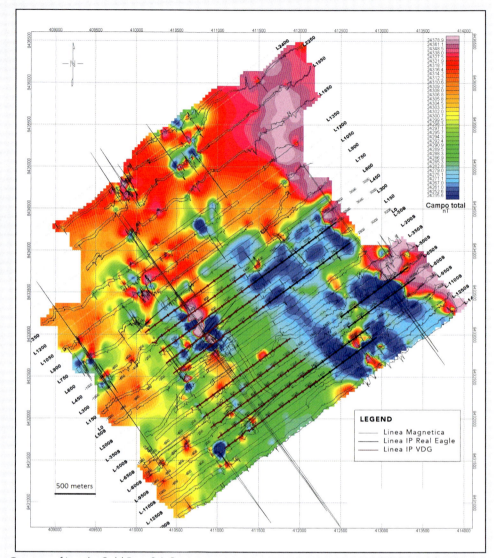

Courtesy of Lupaka Gold Peru S.A.C.

Mineral deposits are rare, and thus geophysical measurements are generally quite "dull," even when presented in color. Figure 1.24 is an exception.

Geochemical Surveys

During formation of an ore body, traces of metals will be dispersed in the surrounding rock and soil and in the groundwater percolating through fractures. Exposure of the ore body to air and water causes weathering, mineral alteration, and erosion leading to further dispersion of metals by physical transport or in groundwater flowing over or through an ore body. Trees and plants will absorb the water causing metals to concentrate in their branches and leaves. As a result, anomalous

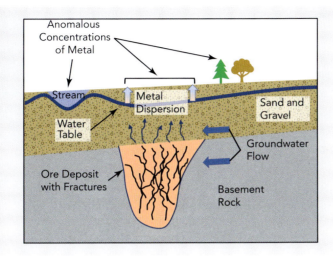

FIGURE 1.25

Formation of a geochemical anomaly near a mineral deposit

concentrations of metals can be detected by taking soil, stream sediment, or plant samples and analyzing them chemically (Figure 1.25). The metal concentrations are measured in parts per million and sometimes parts per billion. If enough samples are taken over a broad area, the measured concentrations may form a geochemical anomaly—higher metal concentrations than background concentrations in the area—that is a target for further exploration.

Generally, concentrations of dispersed metals decrease with distance from the ore body to form a halo around it. Some metals are commonly associated with particular kinds of ore deposits and can be found in a halo at some distance from the ore body, depending on their mobility in soil or water. Such metals are sometimes called "pathfinder elements" (Haldar 2013). An example is the association of arsenic with gold deposits. Arsenic is more mobile than gold and it is relatively easy to measure its concentration. Following the path of increasing arsenic concentrations might lead to a gold deposit.

Drilling

The purpose of drilling is to obtain physical core samples of the rock mass that can be analyzed for mineral or metal concentration. *Core drilling* is done by means of a core barrel, an approximately 3-m-long tube that is inserted into a *drill pipe* (Figure 1.26). At the end of the drill pipe is a diamond drill bit that grinds up the rock as the pipe is rotated and advanced down a hole, forming a cylindrical stub of rock that breaks off and enters the core barrel. A "catcher" on the end of the core barrel keeps the core in the barrel when the barrel is pulled up the pipe.

When the core barrel is full, it is pulled up the drill pipe, and the core is extruded and placed in special boxes. Another length of drill pipe is attached at the top, the empty core barrel is reinserted, and the process continues.

FIGURE 1.26

Core drilling equipment and drill core samples

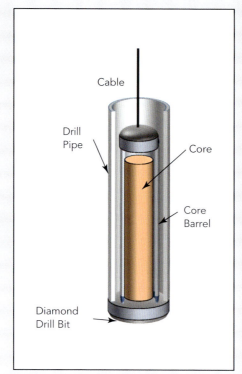

Drill pipe, core barrel, and core

Diamond drill bits (Source: M. Wai, CC BY-SA 3.0)

Drill core samples sawed into halves and quarters (Courtesy of Geology for Investors)

Drill core samples to be logged
(Courtesy of Commerce Resources Corporation and Canadian International Minerals Inc.)

A typical drilling campaign can result in hundreds of meters of drill core samples. Some of the samples are sawed into halves or quarters; one part remains in the field and another part is sent to an assaying lab for chemical analysis or detailed mineralogical analysis using a microscope or X-ray methods.

Geologists look at each of the samples to determine whether there are any economic minerals of interest, or indicators such as minerals or mineral alterations that might occur in an ore-forming system when minerals of economic interest are precipitated. This investigation process is called *core logging*. Figure 1.26 also shows several boxes of drill core samples to be logged.

Diamond core drilling is time-consuming and expensive, but the location of a core sample can be accurately determined, which means that the spatial distribution of metal concentration is accurate. Reverse circulation drilling is a cheaper method and is used to check the results of core drilling. A reverse circulation drill (Figure 1.27) typically uses a piston hammer and a tungsten-steel bit at the end of a double-walled drill pipe to grind up the rock as the pipe advances. Air is forced down the annulus of the drill pipe and forces the rock chips up the pipe to be collected in a bin. The chips are sampled and analyzed. It is assumed that the source of the chips is from the current depth of the drill pipe, though, this is not always the case and thus some inaccuracy results. However, it is usually not significant and the samples from a reverse circulation drill are sufficient for verifying the results of diamond core drilling.

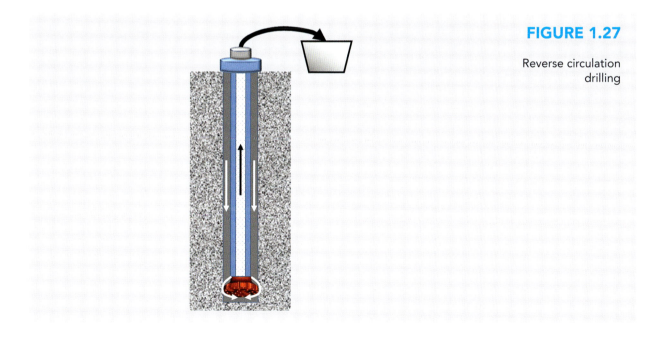

FIGURE 1.27

Reverse circulation drilling

A drilling campaign can cost millions of dollars. Therefore, it is done only when there is enough evidence that economic mineralization might be present. Diamond drilling may cost US$35–$120 per meter whereas reverse circulation drilling may cost US$18–$60 per meter. These are the basic drilling costs. Additional costs are incurred for the mobilization and demobilization of equipment, camp construction and supply, support vehicles, communications, site preparation and drill moves, and supervision. Depending on the contract terms, property location, time of year, and so forth, the additional costs might double the cost per meter.

The goal of drilling is to define an ore-body model. An example of a three-dimensional ore-body model is shown in Figure 1.28 where the samples from a large number of drill holes, indicated by the white lines in the model, had high enough concentrations of metal to define a zone of ore suitable for extraction, indicated by the colored regions. The ore-body model is the basis for design of the mine, whereas the samples of core in the ore zone are used to design the ore processing system.

FIGURE 1.28

Drilling to define an ore-body model. The results of assays of drill cores are used to develop a three-dimensional model of the ore body in which the colors denote grade or some other property of interest.

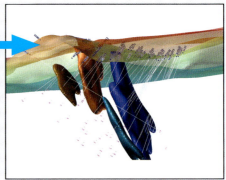

Courtesy of Aben Resources Ltd.   Courtesy of Orefind, www.orefind.com

The drill holes in Figure 1.28 were drilled at an angle to obtain the maximum possible intersection with the ore body. The decision to drill at such an angle would be made by geologists based on information concerning local geological structures.

Mine planning and design are done using a *block model* of the ore body in which the concentrations of metal (grades) derived from chemical analysis (assays) of the drill cores are used to define blocks in which the grade is assumed constant. An example of a block model for a proposed open pit uranium mine is shown in Figure 1.29. The number and size of the blocks depends on the amount of data available. As more data become available, the resolution of the model is improved.

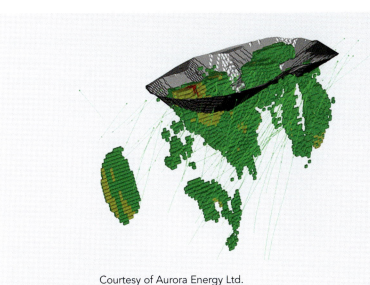

Courtesy of Aurora Energy Ltd.

FIGURE 1.29

Block model of proposed open pit uranium mine

Given a block model, a mining sequence of ore blocks can be devised that maximizes the revenue obtained. To mine these ore blocks, overlying blocks with zero or low grade (waste) must be extracted. The block model can also be used to devise a plan for dealing with waste blocks to minimize any detrimental effects of waste disposal (e.g., acid rock drainage, discussed in Chapter 5).

Block models can also be made for underground mines. In addition, block models can be made for other properties of the ore such as hardness, rock strength, or concentrations of impurities.

IT'S A PIECE OF CAKE

One of the best explanations of the challenges of mine exploration was given by Robert Horn (2015).

Imagine you want to determine the amount of sugar in a cake. Knowing or assuming that sugar is evenly distributed in the cake, you take a small sample and analyze that to determine its sugar concentration, which should be same as the concentration in the cake. This is analogous to finding the concentration of metal in a disseminated deposit—the assumption is that the metal is evenly distributed and that there has been no geological disturbance to alter the distribution.

The problem of finding the number of berries in a muffin is quite different. You sample the muffin and you may or may not hit a berry. If you do find some berries, what assumption do you make about the distribution of berries in the muffin? If you don't find berries, do you assume that there are no berries and return the muffin to the store? This is almost exactly the same as trying to determine the concentration of very small gold particles in a large rock mass.

REFERENCES

Bouzari, F., Hart, C.J.R., Barker, S., and Bissig, T. 2010. Porphyry indicator minerals (PIMs): Exploration for concealed deposits in south central British Columbia (NTS 092I/06, 093A/12, 093N/01, /14). In *Geoscience BC Summary of Activities 2009, Geoscience BC, Report 2010-1*. Vancouver, Canada: Geoscience BC. pp. 25–32.

Galley, A.G., Hannington, M.D., and Jonasson, I.R. 2007. Volcanogenic massive sulphide deposits. In *Mineral Deposits of Canada: A Synthesis of Major Deposit-Types, District Metallogeny, the Evolution of Geological Provinces, and Exploration Methods*. Special Publication 5. Edited by W.D. Goodfellow. St. John's, NL: Geological Association of Canada, Mineral Deposits Division. pp. 141–161.

Haldar, S.K. 2013. *Mineral Exploration: Principles and Applications*. Amsterdam, The Netherlands: Elsevier.

Hazen, R.M., and Ferry, J.M. 2010. Mineral evolution: Mineralogy in the fourth dimension. *Elements* 6:9–12.

Horn, R. 2015. *It's a Minefield: A Prejudiced View of the Mining Industry from the Inside*. Victoria, BC, Canada: Friesen Press.

Images-of-Elements. 2014. Copper, silicon, and gold. https://images-of-elements.com/. Accessed November 2023.

Lee, C.-T.A., Shen, B., Slotnick, B.S., et al. 2012. Continental arc–island arc fluctuations, growth of crustal carbonates, and long-term climate change. *Geosphere* 9(1):21–36.

Öztas, A. 2014. Rose quartz. https://commons.wikimedia.org/wiki/File:Rose_quartz_-_2340.jpg. Accessed August 2023.

Pingstone, A. 2002. Chalcopyrite owl on a base of quartz crystals. https://commons.wikimedia.org/w/index.php?title=File:Chalcopyrite.owl.arp.600pix.jpg&oldid=657677777. Accessed October 2023.

Richards, J.P. 2013. Giant ore deposits formed by optimal alignments and combinations of geological processes. *Nature Geoscience* 6:911–916.

Sepp, S. 2005a. Gneiss. https://commons.wikimedia.org/w/index.php?title=File:Gneiss.jpg&oldid=679469849. Accessed October 2023.

Sepp, S. 2005b. Quartzite. https://commons.wikimedia.org/w/index.php?title=File:Quartzite.jpg&oldid=575714058. Accessed August 2023.

Skinner, H.C.W. 2005. Biominerals. *Mineralogical Magazine* 69(5):621–641.

Stanley, A. 2006. Mount Rundle, Banff, Canada. https://commons.wikimedia.org/w/index.php?title=File:Mount_Rundle,_Banff,_Canada_(200544945).jpg&oldid=724700409. Accessed August 2023.

Tosdal, R.M., Dilles, J.H., and Cooke, D.R. 2009. From source to sinks in auriferous magmatic-hydrothermal porphyry and epithermal deposits. *Elements* 5:289–295.

Wai, M. 2007. Atlas Copco core bits. https://en.wikipedia.org/wiki/File:Atlas_Copco_core_bits.jpg. Accessed November 2023.

Wilson, M.A. 2008. Middle Triassic marginal marine sequence, southwestern Utah. https://en.wikipedia.org/wiki/Sedimentary_rock#/media/File:Triassic_Utah.JPG. Accessed October 2023.

Zander. J. 2007. Pyrite. https://commons.wikimedia.org/w/index.php?title=File:Pyrite_Fools_Gold_Macro_1.JPG&oldid=749884095. Accessed August 2023.

Mining Methods

This chapter describes the methods by which rock that has been classified as ore is extracted from the earth. Surface and underground mining methods are the broad classifications of these methods, but there is considerable variation, especially with underground mines.

An open pit mine is a significant engineering and operational feat, and an underground mine is an extremely complex operation involving what are essentially multiple ongoing construction projects and mind-boggling logistics.

What follows takes you beyond the apparent simplicity of what appears to be only a "hole in the ground" and provide an understanding of the basic concepts and terminology associated with mining operations.

SURFACE MINING

Figure 2.1 shows the typical configuration of an open pit mine (also known as an *open-cut* or *open-cast* mine). Two products come from the pit: ore, which contains the commodity of interest; and waste rock, which is zero- or very-low-grade rock that must be removed to obtain access to the ore. The ore is sent to the processing

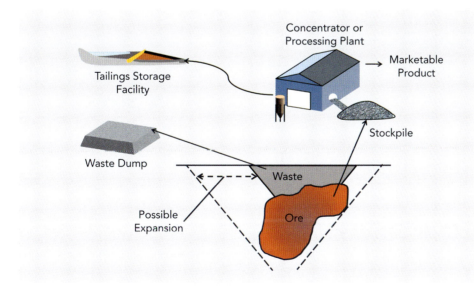

FIGURE 2.1

Open pit mining showing ore, waste, and product streams

plant that produces a marketable product (e.g., a metal or metal concentrate) and tailings, a waste product resulting from processing the ore. The waste rock and tailings are stored in facilities designed to be operational and stable for the life of the mine and after the mine closes.

Figure 2.2 shows a somewhat typical open pit, the Betze-Post open pit at the Goldstrike gold mine operation in Nevada (United States).

FIGURE 2.2

Betze-Post open pit mine, Nevada

Courtesy of Barrick Gold Corporation

Why Are There "Steps" on the Pit Slopes?

A pit slope angle is measured from a horizontal line, as shown in Figure 2.3. The average slope of an open pit mine is about 45 degrees. A 45-degree slope is very steep—and it's scary to look down such a slope.

FIGURE 2.3

Benched pit slope

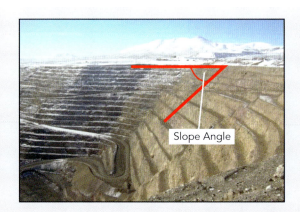

Slope Angle

The stability of the slopes of an open pit mine is a major preoccupation of mine engineers. For safety reasons, it is imperative that the slopes remain standing. This means there should be no rockfalls or large boulders falling from the slope and no rockslides, which are extensive failures involving large quantities of rock. However, one objective of open pit mining is to get to a particular depth as quickly as possible, which means the pit slopes should be as steep as possible to minimize costs and time. But rock slopes cannot be too steep because most rock masses are not strong enough to stand at an angle greater than 40–45 degrees from horizontal. You might think this leads to a typical design problem where one factor is traded off against another. To some extent this is true, but safety and the ability to continue mining are involved and there can be no such trade-offs.

So how do mining engineers deal with pit slope stability? If you drive in the mountains, you will likely see steep rock slopes along the highway. These are stabilized by rock bolts (essentially steel pins), dewatered to relieve water pressures, and covered in steel mesh. Steep mine slopes can be stabilized in the same way, but that is expensive and might not work at all locations. Stabilization work on pit slopes might be done in special situations such as where a haul road or access road is at the base of the slope. However, it is preferable to avoid such stabilization measures.

An important feature of a steep highway slope is a catch area between the highway and the base of the slope where relatively small rockfalls are contained. This provides the fundamental concept for the design of pit slopes in such a way that they can be steep and safe at the same time. The solution is to construct the slope in steps or benches about 5–15 m high, depending on the strength of the rock and the equipment in use. The slopes of the individual benches are steep, usually much greater than 45 degrees, allowing the pit to quickly attain a particular depth, but the overall slope angle is smaller, 30–40 degrees. Slope failures will occur, but their extent will be confined to the immediate bench area and the failed rock mass will be caught by the bench. The concept is shown in Figure 2.4.

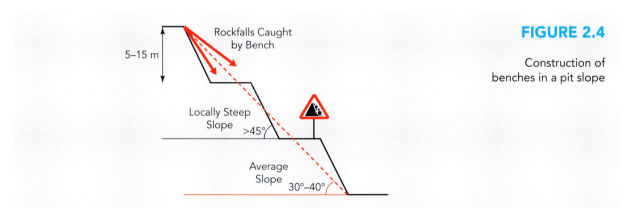

FIGURE 2.4

Construction of benches in a pit slope

A lot of design effort goes into pit slope benches. It's an interesting way of think-ing—to design something that you know is likely to fail, but to design it such that failure has no consequences. Pit slopes are closely monitored using laser reflection from prisms mounted in the slope. If there is any sign of significant movement indicating the possibility of a failure extending across several benches, activity in the pit is stopped until the movements are understood and the pit slope is stabi-lized. Figure 2.5 shows an example of a benched pit slope working as it is supposed to work.

FIGURE 2.5

Benched pit slope at Copper Mountain mine in British Columbia (Canada) showing local slope failures caught by the benches. The benches are 15 m high.

Pushbacks

Maintaining slope angles within stable ranges is important, but it means that if ore is to be extracted from a larger depth, the pit has to be widened to maintain the same slope angle. The geometry is shown in Figure 2.6. The pit widening is called a *pushback*.

FIGURE 2.6

Geometry of a pushback in an open pit mine

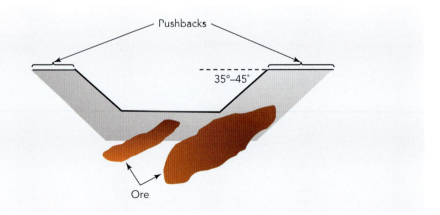

HOW MUCH DOES 5 DEGREES COST?

This requires a bit of geometry and trigonometry, but the bottom line is that it is expensive to make a pit slope less steep. A simple model of an open pit mine is shown below. The volume of the pit is given by this equation:

$$V = \frac{\pi}{3}d(a^2 + ab + b^2)$$

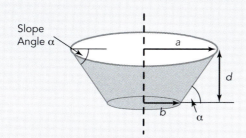

Suppose a = 500 m, b = 250 m, and d = 250 m. Then the total volume of the pit is about 115 million m³ (114,537,232, to be more exact). For these dimensions, the pit slope angle α is 45 degrees.

Now suppose that the rock strength was such that the pit slopes had to be constructed at the shallower angle of 40 degrees. The depth d and the radius of the bottom b are the same, but the radius of the top a increases to about 548 m [500 + 250(1/tan 40 − 1)]. The resulting volume of the pit is larger, about 131 million m³.

The difference in volume of the two pits is 131 − 115 = 16 million m³. This is the extra amount of rock that has to be moved because of the change in slope. How many metric tons is that? The density of the rock might be 2.6 t/m³ so that the difference in volume weighs 42 million t (metric tons). It costs between US$2 and $4 to move 1 t of rock, so the total cost of the change in slope could be as much as 42 × 4 = US$168 million, which is a significant amount, and this is a small pit.

The development of an open pit mine can be considered as a sequence of nested pits, each larger in area than the previous pit. The pushback is the removal of material required to proceed from one pit to the next. Several pushbacks may occur during the life of a mine. The pushback may be in ore or in a combination of ore and waste. The revenue from the ore must pay for the cost of excavating the ore as well as any waste rock.

Figure 2.7 shows the progress of a pushback at the Bagdad mine in Arizona (United States). The pushback began in early 2009 and was complete by 2012. The material removed to construct the pushback was all waste. Ore is being mined from the base of the new pit wall.

FIGURE 2.7

Pushback of the north wall at the Bagdad mine in Arizona

September 23, 2009, looking north   May 15, 2012, looking west

Strip Ratio

To obtain access to ore, some waste rock must be removed. The waste may be a pocket of zero- or low-grade ore within the ore mass or it may be the material removed as part of a pushback. The strip ratio is the amount of waste that must be removed per unit mass of ore:

$$\text{strip ratio} = \frac{\text{waste}}{\text{ore}}$$

Some examples of the strip ratios at different mines are given in Table 2.1. These ratios change during the mine life; for example, they increase during a pushback phase. However, typically, they are large at the beginning of a mine operation and decrease to a "life-of-mine" strip ratio toward the end of the mine life.

TABLE 2.1

Strip ratios for various mines

Mine	Strip Ratio
Oil sands mines, Alberta (Canada)	1.0–1.5
Highland Valley, British Columbia (Canada)	0.45
Bagdad mine, Arizona (USA)	1.4
Red Dog mine, Alaska (USA)	0.8
Cortez mine, Nevada (USA)	2.2

It is important to note that the ore in the equation for the strip ratio is relatively high-grade ore that goes to a concentrator or processing plant. To avoid mistakes when computing the strip ratio, the flow of materials and how they are classified must be determined. In addition to high-grade ore, some mines produce low-grade or oxide ore, which is not processed in a concentrator. In such a case, the total tons must be divided into ore to the concentrator, low-grade or oxide ore, and waste.

SOME STRIP RATIO ALGEBRA

Knowing the total tons mined during a particular time period and the strip ratio, you can figure out how much waste and how much ore are mined. The algebra goes like this:

$$\text{Total mined } T = \text{Ore} + \text{Waste} = O + W$$

$$\text{Strip ratio } S = \frac{W}{O} \Rightarrow W = S \times O$$

$$T = O + S \times O = (1 + S)O$$

$$O = \frac{T}{1 + S}$$

$$W = \frac{S \times T}{1 + S}$$

For example, if $T = 135$ kt/d (kilotons per day), and $S = 0.5$, then the ore mined is $O = 90$ kt/d and the waste mined is $W = 45$ kt/d.

UNDERGROUND MINING

Figure 2.8 shows the typical configuration of an underground mine. Only ore is extracted from the mine. There is a small amount of waste rock generated (development waste) as a result of sinking the shaft or driving the tunnels to gain access to the ore. Sometimes it is used as backfill in the mine. Underground mining results in tailings, the waste resulting from a mineral separation process in the concentrator or processing plant. The significant design issues of an underground mine are geometry of underground mining, ground support, and logistics of materials handling.

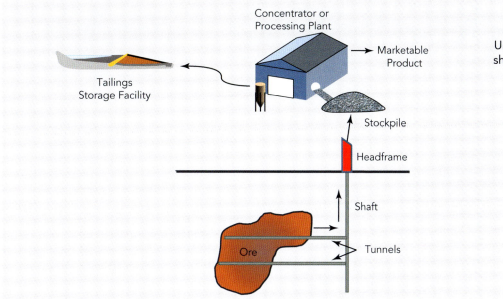

Concentrator or Processing Plant

Marketable Product

Tailings Storage Facility

Stockpile

Headframe

Shaft

Ore

Tunnels

FIGURE 2.8

Underground mine showing waste and ore streams

Underground mines are used to exploit high-grade, deep ore bodies. However, a high-grade, shallow ore body would be mined by underground methods if selectivity were desired, such as in the case of gold or other precious metal vein deposits. For example, the Stillwater platinum mine in Montana (United States) mines very-high-grade ore (12–15 g/t) using a variety of selective underground methods at depths of about 600 m (Sibanye Stillwater Limited 2023).

Surface mines have a limiting depth of 1,000 m (1 km), below which underground mining would always be done. This is because it is difficult to maintain the stability of a 1,000-m-high rock slope and it becomes expensive to haul the ore over large distances to the surface. Two examples of this situation are the Chuquicamata (Chile) and Bingham Canyon (Utah, United States) open pit mines, 850 and 1,200 m deep, respectively. Both mines have begun underground operations.

For underground mines, the mining rate is typically less than 20,000 t/d (metric tons per day); 10,000 t/d is a large capacity (and highly mechanized) underground mine. However, the block caving method, to be described later in this chapter, can achieve mining rates much greater than 20,000 t/d.

There is a considerable amount of terminology associated with underground mining. Some of the more common (but arcane) terms are given in Table 2.2 and illustrated in Figure 2.9.

TABLE 2.2

Underground mining terms

Term	Definition
Crosscut	A horizontal passageway for access to the ore body
Decline or ramp	A spiral or inclined passageway used for access from one level to another
Drift	A horizontal passageway used for access along the length of the ore body
Footwall	The rock below a dipping ore body (this term is not used for a vertical ore body)
Hanging wall	The rock above a dipping ore body (this term is not used for a vertical ore body)
Levels	All the horizontal workings tributary to a shaft station—usually denoted by an elevation from some datum, e.g., the "2300 level"
Orepass	A subvertical chute for movement of ore from one level to another
Raise	A vertical or inclined passageway used for access from one level or drift to another
Stope	An opening from which ore is mined—stopes can be a variety of shapes and sizes
Winze	A vertical internal shaft driven down from an upper level, usually constructed as the beginning stage of a raise

In an underground mine, before personnel and equipment can enter an area, services such as ventilation and power must be available and the rock must be stabilized if necessary. This requires a period of time, often several months or more than

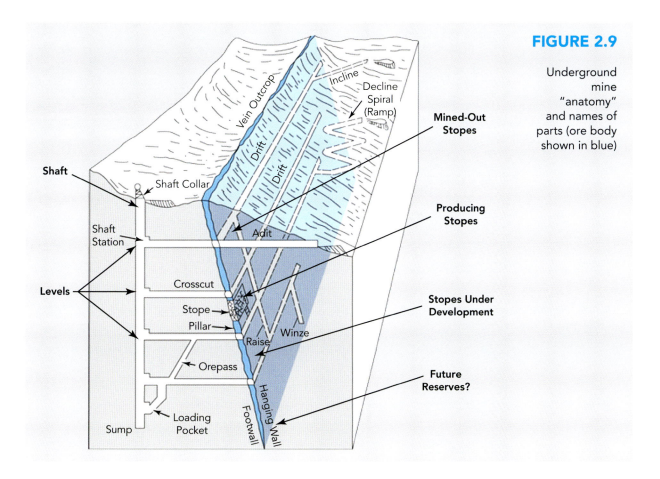

FIGURE 2.9

Underground mine "anatomy" and names of parts (ore body shown in blue)

a year, in which stopes are under development. Also, once a mine is discovered, exploration does not stop. (This is true for both open pit and underground mines.) Thus, in Figure 2.9, in addition to the producing stopes and stopes under development, there are areas where drilling occurs to determine whether sufficient ore is available to justify development to those areas.

Access to an underground mine can be via a decline or ramp; a shaft is constructed for access in deeper mines. Shaft construction is a major undertaking, costing several million dollars. The construction of a shaft, shown in Figure 2.10, is a slow process done in stages by drilling and blasting a vertical cylinder and then hoisting the blasted rock to the surface. Advance rates might be 4–5 m/d (meters per day), but mechanized methods can achieve larger advance rates. In some cases, water in the rock may lead to instability or excessive inflows during construction. To prevent this, the water can be frozen by pumping refrigerants such as brine through vertical pipes to form a stable frozen ring around the shaft before construction.

FIGURE 2.10

Shaft construction at the Lucky Friday mine, Idaho (United States)

Courtesy of Hecla Mining Company

A shaft of diameter 3–6 m is considered small. There are 9- to 10-m-diameter shafts in South Africa (at the Bafokeng Rasimone Platinum and Palabora mines) and a 10-m-diameter shaft in Mongolia at the Oyu Tolgoi mine.

Underground Mining Methods

There is a wide variety of underground mining methods. Several are described and illustrated in the following sections. There are variations of each method. In some mines, more than one method may be used. The decision to use one method over another is based mostly on economics.

Cut-and-Fill Mining

Cut-and-fill mining (Figure 2.11) is applied to steeply dipping ore bodies in stable rock masses. It is a selective mining method and is preferred for ore bodies with irregular shape and scattered mineralization. Because the method involves moving fill material as well as a significant amount of drilling and blasting, it is relatively expensive and therefore done only in high-grade mineralization where selectivity is needed to avoid mining waste or low-grade ore.

Cut-and-fill mining removes ore in horizontal slices, starting from a bottom under-cut and advancing upward. Ore is drilled, blasted, and removed from the stope. When a stope is mined out, the void is backfilled with tailings that are allowed to drain to form a sufficiently solid surface. This is known as *hydraulic fill*. Cement may be added to form *paste backfill*. The fill serves to support the stope walls and provide a working platform for equipment when the next slice is mined.

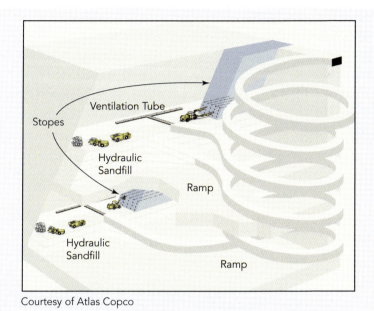

FIGURE 2.11

Cut-and-fill mining

Courtesy of Atlas Copco

Narrow Vein Mining

Narrow vein mining is used in very narrow ore bodies, as small as a half meter wide (Figure 2.12). Blasting is used to "cut" a drift into the rock very close to the vein boundaries. (Blasting can actually surgically break rock.) It is a very selective method; waste rock is left in the hanging wall and the footwall. In a wide vein, a standard load-haul-dump (LHD) machine can operate inside the drift. "Slim-size" machines, including drill rigs, jumbos, and an LHD machine with a 2-m³ bucket, are available for working in drifts as narrow as 2 m. However, in such narrow veins, the use of machines produces waste that dilutes the ore. The alternative is to use a manual technique to extract only the higher-grade material in the vein, but this method is not efficient.

FIGURE 2.12

Mined-out vein of graphite (about 1.2 m wide)

Courtesy of Elcora Advanced Materials

Stoping

Stoping (Figure 2.13) is used for steeply dipping ore deposits in strong rock masses with regular boundaries between the ore and host rock. The ore is recovered from open stopes separated by access drifts. The ore body is divided into sections about 100 m high and laterally into alternating stopes and pillars. A main haulage drive is created in the footwall at the bottom, with cutouts for drawpoints connected to the stopes above. The bottom is V-shaped to funnel the blasted ore (called *muck*) into the drawpoints. Development of the infrastructure for stoping methods is time-consuming, costly, and complex.

FIGURE 2.13

Stoping methods

(Courtesy of Atlas Copco)

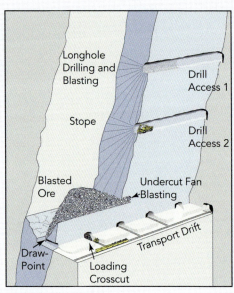

(a) Sublevel stoping   (b) Longhole stoping

In sublevel stoping (Figure 2.13a), short blastholes are drilled from the access drifts in a ring configuration. The ore in the stope is blasted, collected in the drawpoints, and hauled away. The stopes are backfilled with consolidated tailings. This allows for recovery of the ore between the stopes (called *pillars*), enabling a very high recovery of the ore body.

Longhole stoping (Figure 2.13b) uses longer (~100 m) and larger-diameter blast-holes, and thus requires less drilling than sublevel stoping. Greater drilling accuracy is required to avoid including waste in the muck (dilution).

Shrinkage stoping and vertical crater retreat (VCR) are variations of stoping. The VCR method was developed at the Inco (now Vale) mines in Sudbury, Ontario (Canada).

Room-and-Pillar

The room-and-pillar method is typically used in tabular or layered ore bodies such as potash or coal deposits that lie between layers of sedimentary rock. Figure 2.14 shows a pillar in an underground potash mine. The size of the pillars and distance between them is governed by the strength of the rock above and below the ore-body layer. The overlying rock can sag and the underlying rock layer can heave, causing convergence and leading to a decrease in room height. Roof and floor movement and stresses in the pillars are continuously monitored. The ore is typically soft so that the actual mining is done by machines that cut and grind the ore in a continuous fashion.

FIGURE 2.14

A pillar in the Mosaic Company K1 potash mine, Saskatchewan, Canada

Photo © Michael Christopher Brown

Longwall Mining

Longwall mining is a highly mechanized underground mining system for mining coal. A layer of coal is selected and blocked out into an area known as a *panel*. A typical panel might be 3,000 m long by 250 m wide. Passageways are excavated along the length of the panel to provide access and to place a conveying system to transport material out of the mine. Entry tunnels are constructed from the passageways along the width of the panel. A typical configuration of a longwall mining system is illustrated in Figure 2.15.

The longwall system mines between entry tunnels. Extraction is an almost continuous operation involving the use of self-advancing hydraulic roof supports called *shields*, a shearing machine, and a conveyor that runs parallel to the face being

FIGURE 2.15

Configuration of longwall mining in a coal seam

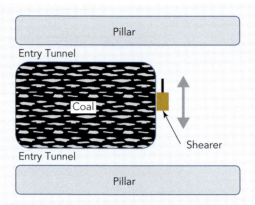

FIGURE 2.16

Shearer in operation in a coal seam

Courtesy of Eickhoff Engine Works and Iron Foundry, Bochum, CC BY-SA 3.0

mined. Depending on rock conditions, the shearer moves along the face at rates between 10 and 30 m/min (meters per minute). Face advance (to the right in Figure 2.16) is about 4–5 m/d.

Block Caving

Block caving is applied to large, low-grade deposits that are too deep to be mined by open pit methods because the strip ratio would be too high. As shown in Figure 2.17, a grid of tunnels, called an *undercut*, is first driven under the ore body. Concurrently or soon after, an underlying system of tunnels and *crosscuts*, collectively called the *production level*, is developed.

Drawpoints are constructed in the crosscuts. Cracking is induced in the ore body by blasting the rock above the undercut. The ore breaks under its own weight forming a cave and flows to the drawpoints. The broken ore is then withdrawn from the

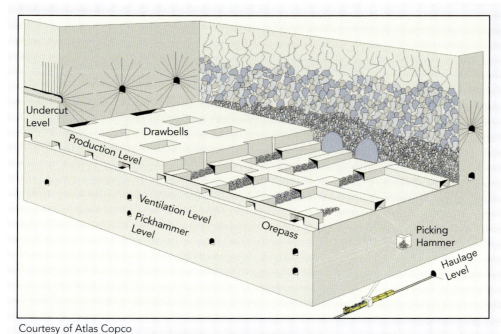

FIGURE 2.17

Block cave mining showing the undercut and production levels

Courtesy of Atlas Copco

drawpoints. As more ore is withdrawn, the cave propagates upward through the ore body until the rock mass overlying the ore body also caves. Surface subsidence might occur.

Caving is induced as a *cave front* across the ore body. Eventually a significant amount of overburden rock enters a group of drawpoints at the cave front, leading to a drop in grade (dilution) below a required minimum value. When this happens, the group of drawpoints is abandoned and the front progresses. Monitoring the grade and performance of the active drawpoints at the cave front is an important aspect of cave mining.

Essentially, block caving creates an underground "inverted open pit" (but without the waste rock). Production rates are high, typically more than 30,000 t/d. It is also amenable to automation, leading to the concept of an "ore factory." Although the per-ton costs of block cave mining are small, the capital costs to develop the underground infrastructure are large and the time required for development is longer than that of other underground methods.

There may be hundreds of drawpoints in a large block cave operation. The withdrawal of ore from these drawpoints must be managed so as not to create significant stresses in the rock mass that could stop the fragmentation process or damage the tunnels in the production level. Once caving in initiated, it cannot be stopped. (Gravity never stops working.) Some proposed block cave operations have 1,000 or more draw-points, which leads to extremely complex technical and management issues.

A list of some operating block cave mines is given in Table 2.3. There is considerable experience with the block cave method in Chile and South Africa. Some copper mine "supercaves" are proposed at Oyu Tolgoi (Mongolia) and at Grasberg (Indonesia). Block caving at a depth of 2 km and a temperature of 80°C is being considered for the Resolution deposit east of Phoenix, Arizona (United States).

TABLE 2.3

Block cave mines currently in operation or planned

Mine	Location	Metric Tons per Day
El Teniente—copper 7 separate operations	South of Santiago, Chile	150,000
Andina—copper and molybdenum	50 km northeast of Santiago, Chile	35,000
New Afton—copper and gold	350 km northeast of Vancouver, Canada	13,000
Northparkes—copper and gold	400 km northwest of Sydney, Australia	18,000
Henderson—molybdenum	70 km west of Denver, Colorado (USA)	36,000
Palabora—copper	360 km northeast of Pretoria, South Africa	30,000
Oyu Tolgoi—copper and gold	South Central Mongolia	90,000 (planned)
Grasberg—copper and gold DMLZ and GBC zones*	Papua Province, Indonesia	200,000

* DMLZ = deep mill level zone; GBC = Grasberg block cave.

Panel caving is a type of caving method in which a very large ore body is divided into areas (panels). As one end of a panel is being undercut, the other end is producing. Production proceeds in a progressive manner across the ore body. Block caving produces from the entire ore body at one time.

ADDITIONAL RESOURCES

For some serious reading about underground mining methods, see SME's *Techniques in Underground Mining: Selections from Underground Mining Methods Handbook*, published in 1998 by Richard E. Gertsch and Richard L. Bullock.

The underground methods described in this chapter can be difficult to understand without animated videos of the methods in operation. An online search of "underground mining methods" can provide more than an evening of entertainment that can be shared with family and friends.

Why All These Methods?

The type of underground mining method used depends mostly on the ore-body geometry, the style of mineralization (e.g., vein or massive), and the rock type and strength. Given these constraints, there is often not much choice. However, there is a trade-off between the operating cost of a method, its ability to extract ore, and its ability to respond in the event of changes in conditions, namely, its flexibility. Higher-cost methods can recover more ore and afford more flexibility. Table 2.4 shows a comparison for selected underground mining methods.

Method	Relative Cost	Recovery/Flexibility
Timbered square set	1.0	100%/High
Cut-and-fill	0.6	100%/Moderate
Shrinkage stope	0.5	80%/Moderate
Sublevel stope	0.4	75%/Low
Room-and-pillar	0.3	50%–80%/High
Longwall	0.2	80%/Low
Sublevel caving	0.5	90%/Low
Block caving	0.2	90%/Low

Adapted from Adler and Thompson 2011

TABLE 2.4

Comparison of underground mining methods

MINING OPERATIONS AND EQUIPMENT

The Mining Cycle

Any underground or open pit mining operation can be divided into distinct stages that form a cycle of successive operations, as shown in Figure 2.18. The cycle begins with the drilling of blastholes and proceeds clockwise through loading the blastholes, blasting, ventilation (if underground), load-haul-dump, scaling of loose rock (underground), installation of support (underground), surveying blastholes, and then back to drilling. The drilling and LHD phases are high-cost items on a per-ton basis.

The volume of rock involved in an open pit blast is considerably greater than that in an underground blast. Consequently, drilling and loading blastholes in an open pit mine can take a few days, but perhaps at most 2 days in an underground mine. For the same reason, moving blasted ore in an open pit requires several days but perhaps a day in an underground mine. Ventilation of gases resulting from blasting underground can take as long as an hour. Scaling and installation of support may take 2 days or more.

FIGURE 2.18

Phases of the mining cycle start at drilling and the process proceeds clockwise. The dollar signs denote the high-cost components of the cycle.

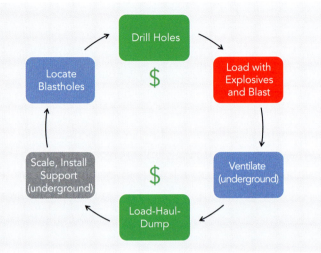

To maintain high utilization of equipment and personnel, it is desirable to have several stopes in operation at the same time, each in different phases of the cycle.

Drills

Drills are used to make holes in rock into which explosives are installed—blastholes. All drilling in hard rock is done with a diamond bit attached to a drill stem, which is a series of pipes connected with threads. The drill stem is rotated to grind the rock in the drill hole. The cuttings resulting from the drilling are sometimes sampled and assayed to provide a control on grade of ore entering the processing plant.

Figure 2.19a shows a remotely operated longhole drill. Figure 2.19b shows a drill jumbo, which is several drills (at least two, sometimes up to five) mounted on one machine and powered by a single-drive system. Water is used as a lubricant and to cool the drills.

FIGURE 2.19

Underground mine drills

(a) Remotely operated longhole rig
(Courtesy of Atlas Copco)

(b) Drill jumbo
(Courtesy of Barrick Gold Corporation)

In an open pit mine, an air rotary drill is used to drill blastholes. An air rotary drill in operation is shown in Figure 2.20a. Air is forced down the drill stem and escapes through small holes in a tricone drill bit shown in Figure 2.20b. The air lifts the cuttings up the drill stem pipe and cools the bit. The cuttings form a pile around the borehole at the surface. Piles of cuttings from drilled blastholes are shown in the foreground of Figure 2.20a.

(a) Air rotary drill

Air

(b) Tricone bit

FIGURE 2.20

Surface mine drills

Blasting

Blastholes are drilled in a pattern as shown in Figure 2.21. The holes are filled with an explosive, typically a combination of ammonium nitrate crystals (called *prills*) and fuel oil (ANFO) if the blastholes are dry. By itself, ANFO is nonexplosive and needs a detonator to start the explosive reaction between the ammonium nitrate and the fuel oil.

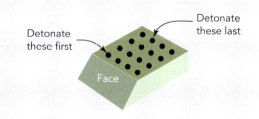

Detonate these first

Detonate these last

Face

FIGURE 2.21

Blasthole pattern showing relative detonation times

Detonation of each row of blastholes is completed in a particular sequence. In Figure 2.21, the row of blastholes closest to the free rock face is detonated first, and then detonation proceeds to the row of blastholes farthest from the face. The delay between the detonations of each row is milliseconds. This sequence avoids blasting too much solid rock—each row has a free face once the preceding row has detonated. (Videos of detonation patterns in open pit mines can be found from a Google search with the keywords "mine blasting videos.")

Ideally, the blast should just lift up the rock mass and fracture it, after which the fractured rock settles. The goal is to contain the gases generated by the blast within the rock mass to fracture the rock as much as possible. In practice, some "venting" will occur before the blast is complete.

Underground blasthole patterns contain fewer blastholes than those of an open pit. They are filled with ANFO, dynamite sticks, or explosive emulsions. Figure 2.22 shows a set of blastholes drilled into a stope face. The explosives and detonators have been installed and are being wired to the detonation system.

FIGURE 2.22

Loaded blastholes being wired for detonation

Courtesy of Newmont Mining Corporation

Ventilation and Refrigeration

A ventilation system is essential to the operation of an underground mine. It must be installed and operational before equipment and personnel can enter the mine. The purpose of ventilation is to dilute and remove various noxious gases that arise from blasting and diesel exhaust as well as from the ore body itself, particularly in the case of coal mines where methane can be a significant emission.

The airflow requirements in the mine are regulated and depend mostly on the number of diesel-powered machines in operation. For example, the Ministry of Labour in Ontario (Canada) requires 0.06 m³/s, or cubic meters per second, for each kilowatt of power of diesel-powered equipment in operation. If an LHD machine has a power rating of 140 kW, an airflow of 0.06 × 140 = 8.6 m³/s is required. If three such machines and some other diesel-powered machine are in operation, then the airflow requirements might be 30 m³/s or more. This is a lot of airflow, more than 2.5 million m³ in a day, and for that reason, large fans and exhaust systems are required, as shown in Figure 2.23.

FIGURE 2.23

Mine exhaust system (inset shows ventilation fan)

Courtesy of AirEng Pty Ltd.

As mine depths increase to 2 km or more, high geothermal gradients (more than the normal 20°C/km) cause rock temperatures to exceed 70°C. The result is a hot mine environment that must be refrigerated for health, safety, and productivity reasons. Refrigeration of the air in a mine uses the same principle as air conditioning. Warm air is transferred from the mine to a heat exchanger containing a refrigerant (e.g., ammonia) that extracts heat from the air and boils off. The cooled air is returned to the mine. The refrigerant vapor is then compressed to high pressure, causing it to condense and release heat. The heat is transferred to air in a heat exchanger and released at the surface. The refrigerant is then returned to the evaporator and the cycle continues. Power requirements and restrictions on the use of certain types of refrigerants are two of the current concerns with mine refrigeration.

Trucks, Shovels, and Scoops

Mining haul trucks and shovels are enormous pieces of equipment and very large capital investments. Figure 2.24 shows a haul truck in operation. It has a capacity

FIGURE 2.24

FIGURE 2.24

Komatsu 830E haul truck at Copper Mountain mine, British Columbia

of 218 t, costs US$5–$6 million, and weighs 154 t when empty. Mines might have fleets of 10 to 60 haul trucks, depending on the size of the operation.

Large shovels serve these haul trucks. Figure 2.25a shows an electric rope shovel that is powered by electricity and Figure 2.25b shows a hydraulic shovel powered by a diesel engine. The shovels shown in Figure 2.25 have bucket capacities between 40 and 110 t. Shovels of this size cost about US$15 million and weigh between 1,000 and 1,500 t. A large mine might have a fleet of four or more shovels.

FIGURE 2.25

Mining shovels

(Courtesy of Caterpillar Inc.)

(a) CAT 7495 electric rope shovel with Hydracrowd

(b) CAT 6040 hydraulic mining shovel

One goal of large machines is to spread fixed costs over a smaller number of larger units, that is, to obtain economies of scale. However, there is concern that maintenance costs of these large machines, particularly tire costs, are too high, leading to diseconomies of scale. Haul trucks with a capacity of 1,000 t have been suggested. Shovels will have to become correspondingly larger—say, 400 m^3. But there are some issues with this idea:

- It may not be possible to build tires for such large trucks.
- New materials and new designs may be needed to build the trucks and shovels.
- There are space constraints on haul roads and maintenance facilities.
- Total production and transportation costs increase with size.
- There are reliability and flexibility issues—if one large machine breaks down, the system stops.

A scooptram, or scoop, or LHD unit, is a diesel or electric-powered machine used to haul ore underground (Figure 2.26). LHD machines have capacities that range from 4 to 22 t. They cost a few hundred thousand dollars, depending on size. Some LHD units have a low profile, between 1.5 and 1.75 m high.

Courtesy of Atlas Copco

FIGURE 2.26

Atlas Copco ST7/ LP scooptram (the driver sits perpendicular to the direction of travel)

Installation of Support

In an underground mine, various rock-strengthening and support systems may have to be installed to provide a safe working environment.

Rock is strong when it is compressed—it takes a huge force to break rock by squeezing it. In the earth, rock is subjected to very large compressive forces because of the weight of rock lying above it. However, opening a hole in rock removes this compressive force and the rock expands. It may be only a few millimeters, but this expansion is enough to cause failures along cracks in the rock. The expansion actually pulls the rock apart, and rock is weak when this happens—it is said to be *weak in tension*.

Thus, cracks form on the roof and walls of an opening and the rock mass begins to fall apart. This might lead to large (dangerous) chunks of rock forming on the roof and walls. When these fall into the opening, it is called *spalling*. Mesh composed of thick wire is attached to the mine roof to prevent this. The mesh is unrolled from a handler installed on one of two booms (Figure 2.27). A drill on the other boom bores holes to install bolts that hold the mesh in place.

FIGURE 2.27

High-tensile steel mesh being installed in mine roof and walls

Courtesy of Geobrugg AG

A compressive force can be imposed on the rock mass around an opening by means of rock bolts (Figure 2.28). A standard rock bolt (Figure 2.28a) is a long steel bar with wedges on one end that locks it into place at the end of a hole drilled from the opening. A nut on the other end of the bolt is tightened to provide compression to the rock mass by squeezing it together. Figure 2.28b shows the Swellex bolt, which is compressed against the sides of the hole by means of water pressure.

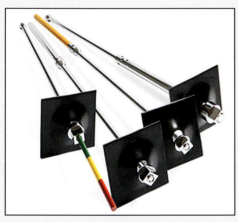

FIGURE 2.28

Types of rock bolts
(Courtesy of Atlas Copco)

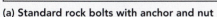

(a) Standard rock bolts with anchor and nut   (b) Swellex rock bolt

Rock bolts provide the same kind of strengthening of a rock mass as reinforced steel rods (rebar) do to concrete. Concrete is also weak in tension, and the presence of steel rods in concrete prevents so-called tensile fractures. The difference is that rock bolts are pressed against the rock mass whereas concrete solidifies around the steel rods.

Shotcrete is a thin layer of concrete that is sprayed onto the rock face (Figure 2.29). It prevents further expansion of the rock mass into the opening and is an alternative to mesh, but it is more expensive. Sometimes short, narrow plastic rods are embedded to provide tensile strength to the shotcrete. What is used for support and strength depends on the rock type and its conditions. It also depends on the use of the opening—for example, permanent or temporary.

Dewatering

Water (groundwater) seeps from the rock mass into mines (both open pit and underground) and it must be pumped out to maintain a dry working area. The quantities of water are usually small and can be relatively easily managed by a drainage system combined with a few pumps.

FIGURE 2.29

Shotcrete installation

Courtesy of Atlas Copco

Some mines penetrate rock masses that have large quantities of water in their fractures and pores (they are saturated). Such rock masses are called *aquifers*. The water in the aquifer may also be under pressure because of its location relative to the aquifer recharge area (e.g., in nearby mountains). This leads to large inflows that cannot be handled by a few pumps and pipes.

One example of this situation is the Cortez mine in Nevada. The mine has 40 high-capacity pumps at locations around the pits. Each pump extracts water out of the ground at a total rate of more than 100,000 L/min to lower the water table and keep the pits dry (Figure 2.30).

Grade Control

The distribution of minerals in an ore body is by no means uniform. Pockets of high and low to zero grade occur within the rock mass. It is desirable to avoid these pockets of waste that lead to *dilution* of the ore. Dilution is distinct from the waste produced during underground development or stripping before open pit mining. Diluted ore containing waste goes to the mill, whereas waste rock goes to the waste dump.

Dilution, *d*, is defined as

$$d = \frac{\text{waste}}{\text{waste} + \text{ore}} \times 100$$

that is, a percentage of the total material extracted. The grade of diluted ore is $g \times (1 - d)$, less than the grade g of undiluted ore. Thus, if dilution is present, more ore

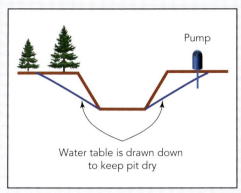

Pump

Water table is drawn down
to keep pit dry

Drawdown of water table next to an open pit

One of 40 pumps located around the Cortez mine in Nevada

FIGURE 2.30

Dewatering of an
open pit mine

YES, THEY STILL USE TIMBER SUPPORTS JUST LIKE IN THE MOVIES!

In the 1860s, a German engineer, Phillip Deidesheimer, developed a modular timber support system that allowed the continuation of operations at the Ophir mine in Virginia City, Nevada (United States). The supports were very large pieces of timber arranged in a cube about 2 m high and 1.5 m wide. Timber supports are still used in some underground mining operations, but the timbers are smaller in size and designed to yield and remain deformed instead of buckling or even snapping catastrophically if subjected to loads caused by ground movement. The photo shows the installation of a 4-m-high wooden support.

Courtesy of Strata Worldwide

must be sent to the mill to obtain the same metal production. Dilution depends on the mining method, ore-body type, and its geometry, as well as the level of grade control and monitoring.

Figure 2.31 illustrates two types of dilution and waste rock in an open pit mine. The pit rim location is somewhat fixed because of the need to maintain pit slope stability. This results in waste rock that goes to the waste dump. Attempts to recover all the ore within the pit will result in *external* dilution where, owing to large machine size, it is impossible to avoid mining some waste. Within the ore body, there are pockets of waste that may be difficult to avoid—again, because of machinery size and also due to the difficulty of separating ore from waste in a pile of muck or blasted rock. These result in *internal* dilution.

FIGURE 2.31

External and internal dilution and waste rock

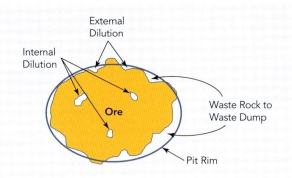

Dilution control (or grade control) is the full-time occupation of a number of people at a mine—geologists, drillers, assay laboratory technicians, and engineers. In open pit operations, GPS systems are used to accurately position individual shovel scoops and drills. Instrumentation on the drill stem is used to sense an ore–waste boundary so that drilling can stop at that point.

In underground operations, assays of the cuttings from blasthole drilling are used to determine whether to fill a blasthole with explosives (Figure 2.32). However, to allow the movement of machinery, it is often necessary to remove some waste that becomes included with the ore. Some current research is being done on methods of separating waste from ore at the source, that is, before it goes up the shaft.

In either open pit or underground operations, there is a trade-off between recovery and dilution. Although it is possible to recover all the ore by simply mining larger volumes, the result will be considerable dilution.

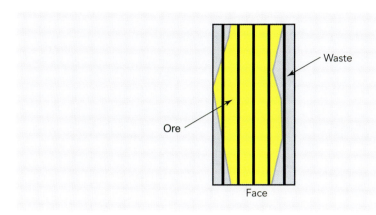

FIGURE 2.32

Blast holes in a stope showing potential for dilution

SOME BIG IDEAS

Mining involves moving large amounts of material and equipment, which is costly. The idea of using large machines to spread fixed costs over a smaller number of larger units was mentioned earlier in this chapter, but there may be physical constraints to such large machines. An interesting alternative is the modular system developed by the European Truck Factory (ETF) in which 240- to 500-t haul trucks are assembled into a haul train, as shown in Figure 2.33.

FIGURE 2.33

Haul train composed of several haul trucks

Courtesy of ETF Mining Trucks

Another transport alternative is dirigibles. Large dirigibles capable of carrying 20- to 50-t loads have been developed by Hybrid Air Vehicles, a U.K.-based company (Figure 2.34). These aircraft could transport equipment and products over sensitive terrain, such as the arctic tundra, thereby avoiding road construction.

FIGURE 2.34

Dirigible depicted flying over an open pit mine

Courtesy of Hybrid Air Vehicles, hybridairvehicles.com

REFERENCES

Adler, L., and Thompson, S.D. 2011. Mining methods classification system (Table 6.2-13). In *SME Mining Engineering Handbook*, 3rd ed. Edited by P. Darling. Englewood, CO: SME.

Eickhoff Engine Works and Iron Foundry, Bochum. 2006. Shearer at work in a coal mine. https://commons.wikimedia.org/w/index.php?title=File:SL500_01 .jpg&oldid=484630537. Accessed August 2023.

Engineering.com. 2006. Syncrude. October 13. www2.engineering.com/Library /ArticlesPage/tabid/85/ArticleID/69/Syncrude.aspx. Accessed October 2023.

Sibanye Stillwater Limited. 2023. Stillwater & East Boulder. www.sibanyestillwater.com /business/americas/pgm-operations-americas/stillwater-east-boulder/. Accessed July 2023.

Mineral Processing and Metal Refining

Mineral processing (also known as *mineral dressing*) is a collection of physical and chemical processes that separate minerals of value from gangue (pronounced "gang"), the minerals with no value. These processes happen in buildings such as those shown in Figure 3.1.

After the gangue is separated from valuable minerals, further processing and refining are necessary to break down the minerals into pure metals, a chemical form of separation.

The same principle applies to nonmetallic ores—waste or unwanted material is separated from the material of interest. Generally, physical methods are used to separate the waste from the material of interest.

Mineral processing and refining are an interesting part of the mining industry. A wide variety of methods are used and a few are described in this chapter. The chemistry in Appendix A is relevant to this chapter, but not essential.

COMMINUTION

For metallic ore, the relationship between valuable minerals and gangue in the ore deposit is illustrated by the drill core shown in Figure 3.2. The valuable minerals are the shiny specks (copper and nickel sulfides in this case) and the gangue is the white-gray silicate minerals (mostly quartz in this photo). Before any chemistry is done to get the metal out of this ore, we need to get physical and break up the ore to allow separation of the shiny stuff from the less shiny stuff, sometimes known as *liberation* of the shiny stuff. (The word *liberation* is appropriate because the dull minerals are more numerous and encircle or trap the shiny ones.)

Comminution is the name of the general process of breaking up solids from one average particle size to a smaller average particle size. In the case of mineral processing, comminution is done by crushing and grinding. Blasting, discussed in Chapter 2, is actually the first stage of comminution.

FIGURE 3.1

Concentrators
and processing
plant at various
mine sites

Concentrator at Copper Mountain mine, British Columbia, Canada
(Courtesy of Copper Mountain Mining Corporation)

Concentrator complex at Oyu Tolgoi mine, Mongolia
(Courtesy of Turquoise Hill Resources)

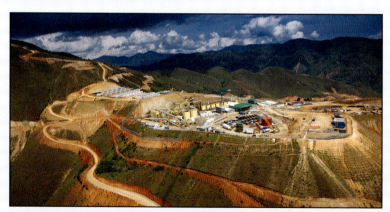

Gold processing plant, Namoya mine, Democratic Republic of Congo
(Courtesy of Banro Corporation)

FIGURE 3.2

Drill core showing valuable minerals (shiny specks) and gangue (white silicates)

Crushing

Crushing is the second stage of comminution. It results in rock particles with sizes ranging between 10 and 50 mm. Three types of crusher equipment are described in the following sections.

Gyratory Crusher

A gyratory crusher is used in high-throughput operations, more than 500 t/h (metric tons per hour). Figure 3.3a shows a cross section of a gyratory crusher. A conical crushing head is attached to a central spindle suspended from a support called a *spider* at the top of the crusher. The spindle moves eccentrically about the vertical axis so that the rock particles are broken down into successively smaller particles between the crushing head and the hardened steel lining of the crusher shell. Figure 3.3b is a photo of the ore feed opening of a gyratory crusher.

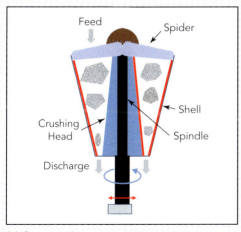

(a) Cross section

(b) Ore feed opening at top
(Photo © Metso Corporation)

FIGURE 3.3

Gyratory crusher

A hydraulic rock breaker might be used to break up large rock fragments that could jam the crusher. Ideally, these large fragments would not be produced by the blasting stage.

Jaw Crusher

A jaw crusher is used in operations that process less than 500 t/h. Figure 3.4 shows a cross section of a jaw crusher. The ore is fed into a tapered crushing chamber formed by a fixed jaw and a moving jaw. Rock particles are crushed to successively smaller sizes as they travel downward. The crushing force is provided by the eccentric motion of a weighted flywheel. The opening at the bottom of the jaws is adjusted to the desired maximum particle size.

FIGURE 3.4

Operation of a jaw crusher

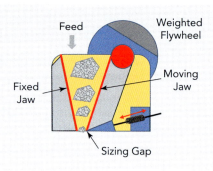

Cone Crusher

A cone crusher works much like a gyratory crusher except that the crushing chamber is flatter and the conical crushing head is supported from below (Figure 3.5). The flatter crushing chamber causes particles to remain in the crusher longer, thus producing finer particles in the discharge. A cone crusher is used as a secondary crusher to break medium-hard to hard rocks that are not broken by either primary crushing or grinding.

FIGURE 3.5

Operation of a cone crusher. The spindle rotates eccentrically, which moves the cone against the mantle. Rock particles are broken into successively smaller sizes as they move toward the outlet.

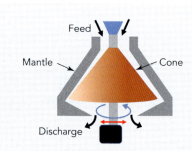

Crushing is often done in the mine, either open pit or underground, after which the crushed ore is transported to the processing plant. Because the rate at which ore is produced in the mine is less than the rate at which ore is processed, crushed ore is stored in a stockpile near the processing plant. Ore stockpiles are shown in Figure 3.6.

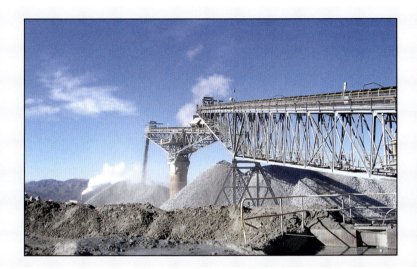

FIGURE 3.6

Crushed ore stockpiles at the Bagdad copper mine in Arizona (United States). The end of the conveyor moves to different stockpile locations as necessary.

The length of time that crushed ore can be left out in the open depends on the climate and the types of ore minerals. Chemical changes of the minerals, such as oxidation, may occur, leaving a coating on the mineral surfaces that could interfere with processing the ore.

Grinding

Grinding is the third stage of comminution. It is done in the processing plant. Crushed ore is mixed with water to form slurry that is then pumped into rotating cylindrical grinding mills. The goal is to reduce particles to sizes less than 1 mm. There are two grinding steps: a coarse grind followed by a fine grind.

The coarse grind is done in either an autogenous grinding (AG) mill or a semiautogenous grinding (SAG) mill. In an AG mill, the large particles of crushed ore are used as the grinding medium; that is, the ore is "self-ground," hence the name *autogenous*. For successful autogenous grinding, the ore must be hard and it must break along boundaries between mineral grains to produce particles large enough to grind the remaining particles to sufficiently fine size. If the ore cannot be ground autogenously to sufficiently fine sizes, semiautogenous grinding is used in which steel balls and the ore itself are tumbled to break the ore.

Whichever type of mill is used, the desired maximum size of particles exiting the mill is 10 mm, or about 0.4 in., although the average size could be much smaller. Figure 3.7 shows a SAG mill at the Highland Valley copper mine in British Columbia (Canada). The diameter of AG and SAG mills is normally two to three times the length.

A combination of impact, attrition, and abrasion cause the particle size reduction in the mill. The inside wall of AG/SAG mills consists of steel or rubber liners and lifter plates that lift the material to the top of the mill from where it falls onto the ore at the bottom, causing impact forces. Particles slide along each other as they move to different heights and are subjected to attrition and abrasion. Figure 3.8 shows a computer simulation of the material motion inside a SAG mill.

SOME PHYSICS OF COMMINUTION

If the peripheral speed of the mill is too great, the mill begins to act like a centrifuge, and the ore particles (and balls) do not fall back into the center of the mill but stay on the perimeter. The rotation speed at which this occurs is called the *critical speed*, measured in revolutions per minute. The larger the diameter, the slower is the critical speed. The critical speed of a 5-m-diameter mill is about 19 rpm (revolutions per minute) whereas that of a 10-m-diameter mill is about 13 rpm. Mills usually operate at 65%–75% of the critical speed.

A large-diameter, slowly rotating mill relies on the ore and balls falling through a large distance to break up the ore, whereas a longer, faster rotating mill relies on a longer residence time.

The largest user of energy at a mine site is the comminution processes. The graph shows the energy use per ton of ore for blasting, crushing, and grinding. Grinding uses the most energy because of the larger size change, from 10 to 0.075 mm or less. For this reason, considerable research is being done to find more efficient grinding methods and to improve the fragmentation caused by blasting.

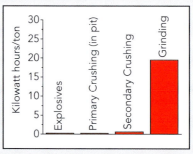

Data from Workman and Eloranta 2003

FIGURE 3.7

SAG mill at the Highland Valley copper mine, British Columbia. The diameter of the mill is about 10 m.

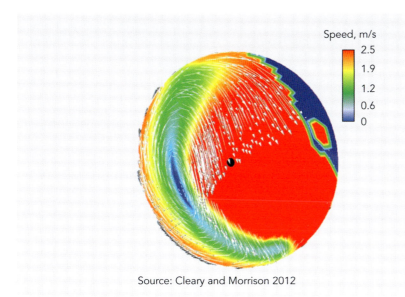

Speed, m/s

2.5
1.9
1.2
0.6
0

Source: Cleary and Morrison 2012

FIGURE 3.8

Computer simulation of the motion of steel balls and material inside a SAG mill. The trajectories of material are shown by white arrows. The red denotes high speed, the green means intermediate, and blue is stationary. The mill is rotating clockwise.

The fine grind is done by a ball mill that, as the name implies, grinds material by rotating a cylinder containing the ore and steel balls, causing the balls to fall onto the ore at the bottom of the cylinder. Size reduction is caused mostly by impact. Figure 3.9 shows a charge of steel balls used in SAG or ball mills. One of the ball mills at the Highland Valley mine is shown in Figure 3.10. The diameter of a ball mill is typically half its length.

FIGURE 3.9

Steel balls used in
SAG or ball mills,
Highland Valley mine,
British Columbia

FIGURE 3.10

Ball mill (foreground),
Highland Valley mine,
British Columbia

Ball mills are used to grind material to a particle size between 20 and 75 μm, or micrometers (0.02–0.075 mm). Some gold ores are ground to particle sizes of 20–30 μm to allow separation of the small gold particles.

Cyclones

An important part of the comminution process is a cyclone (or hydrocyclone). This machine is used to separate coarse-grained particles from fine-grained particles in slurry (a mixture of particles and water). This separation process is also called *classification*.

The slurry feed is pumped tangentially under high pressure (and velocity) through an inlet at the top of the cyclone (Figure 3.11). This generates a centrifugal force large enough to generate a low-pressure air core along the axis of the cyclone. The centrifugal force causes the larger and heavier particles to collect at the cyclone wall where the velocity is lowest and they migrate downward. The motion of smaller and lighter particles is inhibited by the larger particles, and they become caught in the low-pressure zone along the axis and are carried upward. The small particles are called *overflow*, and the large particles are called *underflow*.

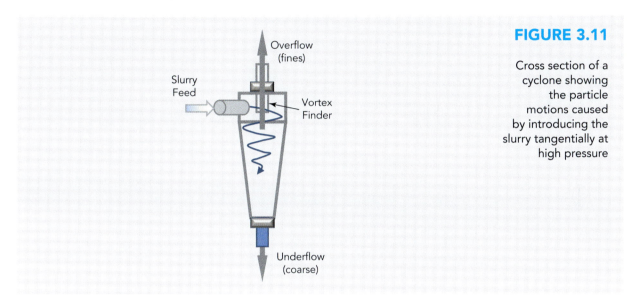

FIGURE 3.11

Cross section of a cyclone showing the particle motions caused by introducing the slurry tangentially at high pressure

Water content of the slurry might be 50% by volume. One might think that water content has little to do with the particle sizes exiting the cyclone, but, in fact, increasing the water content decreases the sizes of the particles in the underflow. The reason is that the more water in the slurry, the less the motion of the smaller and lighter particles is inhibited by the larger particles. Thus, more of the smaller particles are able to migrate to the outside wall where gravitational forces predominate, forcing the particles to the underflow.

Grinding Circuits

Crushers, AG and SAG mills, ball mills, and cyclones are configured into grinding circuits. This is best illustrated by an example. At the Bagdad copper mine in Arizona, five grinding circuits in the mill process 75,000 t of ore per day (3,125 t/h). The configuration of one circuit is shown in Figure 3.12. The output of an AG mill is fed into a screen. The coarse material from the screen is passed to a cone crusher and fed back into the AG mill. One cone crusher is connected to all five circuits. The cone crusher is used to break up larger particles that would

FIGURE 3.12

Grinding circuit at the Bagdad copper mine, Arizona

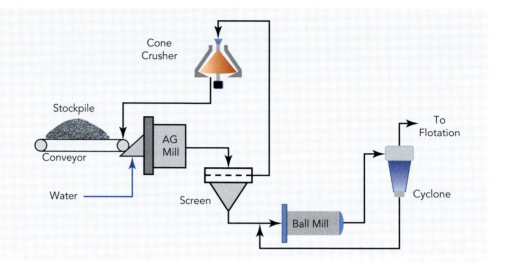

otherwise simply cycle through the AG mills. The fine material from the screen is fed into a ball mill circuit. The output of the ball mill is separated into coarse and fine fractions in a cyclone; the coarse fraction (underflow) is recycled and the fine fraction (overflow) is pumped to the flotation tanks.

Grinding circuits typically involve secondary crushing or regrinding, cycling particles from the output of one unit back to the input of the unit. Recycling particle paths, similar to the two shown in the Bagdad circuit of Figure 3.12, are found in all grinding circuits. These ensure the most consistent feed size range possible without wasting energy in grinding.

The design of grinding circuits is a complex field (some call it an art). The equipment used and the manner in which it is connected depends on the way the ore breaks up into finer sizes, which depends mostly on the hardness of the ore.

One common question is: How long does a particle stay in a grinding circuit before it is ground to a size fine enough for flotation? This is known as the *residence time* for the particles. It depends on a number of factors such as the size of the mills and the ore hardness. However, a typical residence time might be between 5 and 12 minutes.

Flotation

Froth flotation is the most common method for separating sulfide minerals from each other and from waste minerals or gangue. Precursors to the method were developed in the late 19th century, but the basis of the modern process was developed in the early 20th century. Before flotation became common practice, copper, lead, and zinc were obtained from high-grade underground deposits. By the middle of the 20th century, flotation was widely used and enabled copper sulfides to be separated from rock with grades much less than 1% mined from large open pit deposits.

Flotation is considered one of the most significant enabling technologies of the last century. It is used in other applications involving separation, such as de-inking recycled paper and in wastewater treatment.

Figure 3.13 illustrates the flotation process as it is used in the mining industry. The slurry from the grinding circuit is mixed with an organic chemical called a *collector*. It selectively coats, or adsorbs onto, the surface of the sulfide mineral of interest and renders it hydrophobic, meaning literally "afraid of water." The slurry (now called a *pulp*) is then pumped into the first of a series of flotation cells.

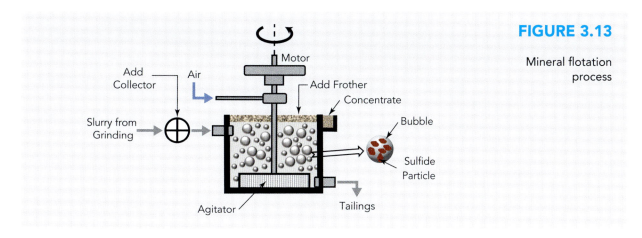

FIGURE 3.13

Mineral flotation process

The pulp is stirred as air is pumped into it from the top. This forms bubbles in the pulp that collide with the sulfide particles. The hydrophobic sulfide particles collide with and attach to the rising bubbles that float to the surface where they collect in a froth layer that flows over the top of the cell into a channel. A *frother*, which is a type of alcohol, is added to stabilize the froth layer.

By way of comparison, the froth on the top of a mug of beer is not stable enough to be used in sulfide flotation.

The result is a wet concentrate at the top of the flotation cell (Figure 3.14). The concentration of the copper in a copper concentrate is between 20% and 30%, considerably greater than the concentration of copper in the mine, about 0.5%. The concentrate flows over a weir into a channel and is sent to further processing or

FIGURE 3.14

Copper concentrate in a flotation cell at the Highland Valley mine, British Columbia

to be dried. The gangue minerals collect at the bottom of the flotation cell and are sent to the tailings storage facilities.

Froth flotation is effective for particle sizes between roughly 40 and 200 μm. Particles larger than 200 μm are too heavy for the bubbles to lift, whereas particles smaller than 40 μm often will not attach to an air bubble.

The first application of flotation to separate sulfides occurred in 1896 at the Broken Hill mine in Australia where eucalyptus oil was used as a collector. Collector chemistry has advanced considerably since then so that different metal sulfides in an ore can be sequentially floated by the use of different types of collectors and adjustment of the flotation cell chemistry.

Flotation Circuits

Up to 12 flotation cells are connected in series to form banks and the banks are connected into circuits. The rationale for these circuits is somewhat similar to that of grinding circuits—namely, recycle the processed material until the properties of the end product become as close to the desired properties as possible. In a flotation circuit, the goal is to recover as much of the metal as possible without diluting the concentrate with unwanted minerals.

A simple circuit is shown in Figure 3.15. Collector and frother are added to the pulp that is then pumped into the first set of flotation tanks called the *rougher cells*. The liberated target mineral particles will float more rapidly than other minerals and therefore the rougher cells produce the concentrate product that is sent for further processing or to be dried. The tailings from the rougher cells are then sent to

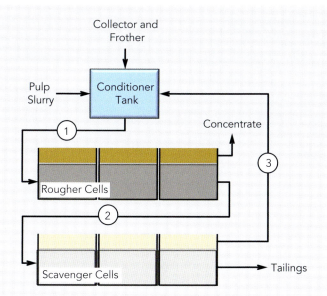

FIGURE 3.15

Flotation circuit

scavenger cells that, as the name implies, obtain as much of the remaining sulfide as possible. The concentrate produced by the scavenger cells is returned to the rougher cells where, once more, collector and frother are added.

The scavenger tailings are usually barren enough to be sent to the tailings storage facility, but in some cases may be sent to cleaner cells to be re-floated. More complex flotation circuits have several sets of rougher, scavenger, cleaner, and re-cleaner cells, as well as intermediate regrinding of pulp or concentrate. Figure 3.16 shows an arrangement of banks of different types of flotation cells at a zinc processing plant in Europe. The yellow housings are the stirrer motors of each flotation cell.

HOUSEHOLD COLLECTORS

A common household collector is soap, which coats dirt particles, but works differently from the collectors used in flotation. One end of a soap molecule is hydrophobic while the other is hydrophilic, meaning "water-loving." Dirt particles are typically hydrophobic. To minimize exposure to water (which they are "afraid of"), the hydrophobic end of the soap molecule will attach to the hydrophobic dirt particle. This leaves the hydrophilic end of the soap molecule in the water (where it wants to be). As the water flows out the drain, so do the soap molecules attached to the dirt particles. And you or your laundry are clean!

A social analogy might be as follows: Dirt and water are having a nice party with lots of interaction and then soap comes in like a loud, attention-seeking guest who causes the party to divide into groups.

FIGURE 3.16

Flotation cell banks at a zinc flotation plant in Europe

Courtesy of Westpro Machinery

The residence time of a particle in a flotation cell depends on the size of the cell and the desired time for the particles to be exposed to the collector. Depending on the number of cells, the residence time in a flotation circuit ranges from 5 to 15 minutes (Gupta and Yan 2016).

A simple materials balance exercise can be used to determine the amount of ore needed to produce 1 ton of concentrate. This is known as the *concentration factor*, or CF. The relationship between what comes into the concentrator and what comes out of it is illustrated in Figure 3.17.

FIGURE 3.17

Material balance between ore body and concentrator

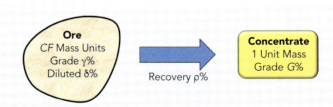

Balancing what goes into the concentrator with what comes out gives this equation:

$$CF \times \gamma \times \rho \times (1 - \delta) = 1 \times G$$

What goes in   must come out

$$CF = \frac{G}{\gamma \times \rho \times (1 - \delta)}$$

Thus, if the ore grade is 0.25% Cu, the recovery of copper in the concentrator is 80%, there is zero dilution, and the concentrate is G = 28% copper, CF = 140 t.

There is an upper limit to the concentration of a metal in a concentrate, depending on the mineral in the ore. This is the direct proportion by atomic weight of the metal to the molecular weight of the mineral. Some approximate atomic weights are given as follows:

Copper	Iron	Lead	Zinc	Sulfur
64	56	207	65	32

For a copper concentrate made from chalcopyrite ($CuFeS_2$), the copper concentration limit is 34.8%, that is, $64/(64 + 56 + 2 \times 32) = 0.348$. Similarly, the concentration limit of lead in a lead concentrate made from galena (PbS) is about 87%, and for a zinc concentrate made from sphalerite (ZnS), the concentration limit is about 67%. A mine that has bornite (Cu_5FeS_4) in its ore can achieve quite high copper concentrations; unfortunately, bornite is relatively rare.

Mineral Separation in Flotation

Separation of minerals by flotation varies depending on the complexity of the ore. Recovery of metal from the ore depends on separation. For a simple ore containing only copper with some gold by-product, recovery can be 90%–95%. Recovery is lower for polymetallic ores that may contain roughly equal proportions of desirable metals. The lower recovery is the result of poor mineral separation that is often caused by the lack of selectivity of the collectors, interaction between different metal sulfides, and complexity of the ore mineralogy.

For example, in addition to copper sulfides such as chalcopyrite ($CuFeS_2$) and bornite (Cu_5FeS_4), a copper ore body may also contain a variety of other metal sulfides such as pyrite (FeS_2), arsenopyrite (FeAsS), and minor amounts of galena (PbS), sphalerite (ZnS), and molybdenite (MoS_2). All these sulfides would be present in the ore slurry and, depending on conditions in the flotation tank, may or may not find their way into the copper concentrate. The presence of pyrite and arsenopyrite lowers the grade of the concentrate, and anything with arsenic in it is considered a contaminant by a copper smelter. Lead and zinc are also considered contaminants by copper smelters, but the grades may be high enough to be of commercial value and therefore warrant recovery from the tailings of the copper flotation. The metals antimony, bismuth, and mercury may also be present in any one of the sulfide minerals but are considered contaminants in a copper smelter.

Separation of all these sulfides is quite a delicate exercise in flotation cell chemistry and may not always be possible using flotation techniques. Flores et al. (2020) give an interesting description of the challenges associated with producing clean, high-grade copper concentrates from ores with complex combinations of sulfide minerals.

Pyrite and arsenopyrite, which are almost always present, are depressed (the opposite of floated) by the addition of lime to the flotation tank, which increases the alkalinity of the tank water. Increased alkalinity means an increased amount of the hydroxyl ion OH^- that forms metal hydroxide films (e.g., FeOH) on the surfaces of pyrite and arsenopyrite and reduces the ability of the collector to adsorb onto these minerals. The minerals sink to the bottom of the tank while the copper minerals, whose surfaces are not affected by the hydroxyl ion, are coated with collector, attach to bubbles and float to the surface.

USES OF MOLYBDENITE AND PYRITE

Molybdenite may be purified for use in lubricants or sent to roasters to form molybdic oxide (MoO_3), which is used to form a hard steel alloy that can withstand high temperatures. Such alloys are used in making high-speed cutting tools, aircraft and missile parts, and forged automobile parts. Other useful molybdenum compounds include ammonium molybdate [$(NH_4)_2MoO_4$], used in chemical analysis for phosphates, and lead molybdate ($PbMoO_4$), used as a pigment in ceramic glazes.

Pyrite has rather limited uses, but it is a semiconductor and some scientists believe it could be used in photovoltaic cells. Another possible use is as a source of iron when primary iron ores, such as hematite and magnetite, become exhausted sometime in the future. It is a very abundant mineral found in many different geological environments.

Molybdenite can be separated relatively easily from chalcopyrite and other copper sulfides by transferring the concentrate to a vat containing sodium hydrosulfide (NaHS), which strips the collector off all sulfide particles. The copper sulfide particles sink to the bottom of the vat where they are collected and sent to a drying facility while the molybdenite rises to the top of the vat as it floats naturally, being a somewhat greasy (hydrophobic) mineral. The molybdenite concentrate formed (containing about 58% molybdenum) is then dried.

The metals of interest may be found in several mineral species, each one responding differently to a particular collector. This is a form of mineralogical complexity. One simple example is lead-zinc ore that contains varying quantities of the minerals galena and sphalerite that are to be separated to produce a lead concentrate and a zinc concentrate. The lead sulfide galena can be floated by commonly used collectors whereas the zinc sulfide sphalerite cannot. A two-stage selective flotation process is therefore used in which galena is first floated to form a lead concentrate and the sphalerite (and any pyrite) is depressed by adding lime. In the second stage, the tailings from the first stage are treated with copper sulfate ($CuSO_4$), which produces the copper ion Cu^{2+} in solution. The copper ion reacts with the sphalerite to produce a film of covellite (CuS) on the surface of the sphalerite particles as follows:

$$ZnS + Cu^{2+} \leftrightarrow CuS + Zn^{2+}$$

where the double-pointed arrow means the reaction can proceed in both directions. The sphalerite is said to be *activated* because the covellite can adsorb a commonly used collector for copper sulfides, allowing flotation to produce a zinc concentrate.

Ore bodies containing nickel-copper-cobalt sulfides are another example of complex ore mineralogy requiring multiple-stage flotation and different reagents to produce separate concentrates. One sulfide concentrate can often be contaminated by other sulfides. For example, the nickel concentration in a copper concentrate derived from a nickel-copper-cobalt ore body must contain less than 0.5% nickel to provide an acceptably pure copper product in the copper refinery. Several methods for treating these ores have been developed.

Precious metals are often included in base metal concentrates, and it is desirable to increase their concentration in the concentrate as much as possible because they can be recovered in the refining process and result in a credit to the mining operation. Figure 3.18 shows a photomicrograph of a rock sample from an ore body in Brazil. Gold (Au) is seen as an inclusion in chalcopyrite crystals (denoted Cpy), which are

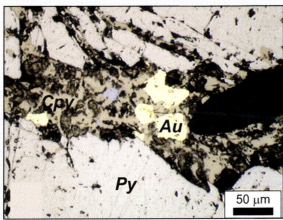

Source: Monteiro et al. 2008; used with permission from Sprott Global Resource Investments Ltd.

FIGURE 3.18

Reflected light photomicrograph of a rock sample from an ore body (Cpy denotes chalcopyrite, Py denotes pyrite, Au denotes gold)

surrounded by pyrite crystals (denoted Py). In this case, the gold is found in the copper concentrate. If the gold occurred as inclusions in the pyrite, then it might be preferable to not depress the pyrite during flotation if the resulting lower concentrate grade could be compensated by the credit for the gold.

The Grade–Recovery Battle

The flow rate and size of a flotation tank are designed to give the minerals enough time to be coated with collector (commonly called *activation*). As the input flow rate decreases, the particles of the sulfide of interest become more exposed to the collector and adhere to the bubbles so that recovery of the sulfide increases.

FIGURE 3.19

Mineral particles in a sulfide ore slurry

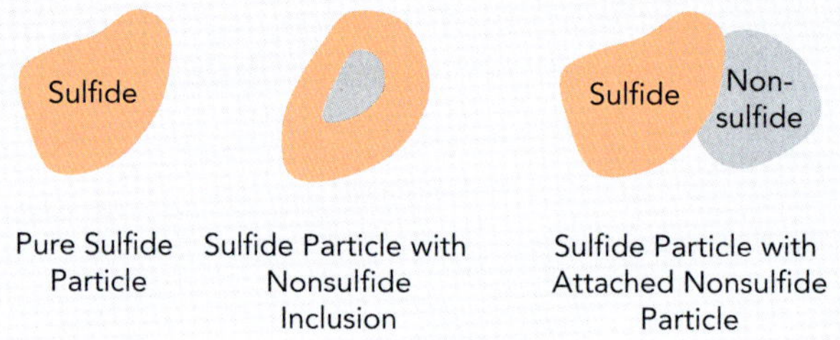

Sulfide	Sulfide	Sulfide	Non-sulfide
Pure Sulfide Particle	Sulfide Particle with Nonsulfide Inclusion	Sulfide Particle with Attached Nonsulfide Particle	

However, the concentrate grade decreases. This is illustrated in Figure 3.19. Particles in sulfide ore slurry could be one of three types: pure sulfide, sulfide with a nonsulfide inclusion, and sulfide with an attached nonsulfide mineral. If the recovery of sulfide minerals of interest is increased, more nonsulfides will be included in the concentrate, which decreases the concentrate grade.

A large amount of monitoring and control of the flotation process is devoted to obtaining the maximum recovery for a given concentrate grade. This is a common problem in all sulfide concentration processes. Another approach is to use finer grinding to try to obtain particles consisting of only one mineral. However, this can be costly and would only be done if the cost could be justified by the recovery of valuable metals.

The grade–recovery trade-off is analogous to the ore recovery problem in a mine where it is possible to recover all the ore but at the expense of considerable ore dilution.

SMELTING AND REFINING

A sulfide concentrate is not the pure metal required—it is a sulfide with a high concentration of the metal of interest. To get the metal of interest, the sulfide has to be broken down (actually burned), and that is done with lots of oxygen and heat. This releases the metal in almost pure form. It is then refined, resulting in pure (typically 99.99%) metal.

Smelting

Smelting is the process used to oxidize the sulfides. The concentrate contains a sulfide mineral that can be oxidized by hot oxygen. More precisely, it is the sulfur atom in the sulfide that is oxidized. Metal sulfides break down naturally when exposed to air. The application of a flow of hot oxygen speeds up this process.

The smelting process for copper concentrate starts in a flash furnace, as shown in Figure 3.20. Copper concentrate (up to 4,500 t/d [metric tons per day]) and silica (in the form of sand) are transferred into the furnace where they are heated by hot oxygen to a temperature of 1,200°C. At this temperature the concentrate melts to form iron oxide, sulfur dioxide, iron sulfide, and chalcocite, a copper sulfide.

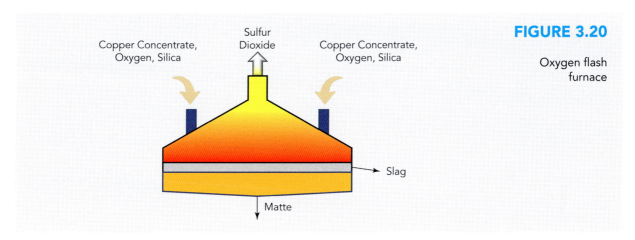

FIGURE 3.20

Oxygen flash furnace

The entire smelting process is illustrated in Figure 3.21. The iron sulfide and the chalcocite are liquids called *matte*. The iron oxide reacts with the silica to form a *slag*. The slag is lighter than the matte so it floats on top; from there it is removed and taken to a disposal site. The matte is tapped off and burned in a converter furnace with more silica to remove iron sulfides, resulting in more slag and *blister* copper, so named because of its blister-like surface caused by the release of sulfur dioxide during cooling. Recycled copper, if available, might be added to the converter.

Excess oxygen in the blister copper is burned off using natural gas, after which the molten blister is molded into an anode, a rectangular solid about 75 mm thick and weighing about 1 t with two hangers at the top.

FIGURE 3.21

Smelting of copper
concentrate

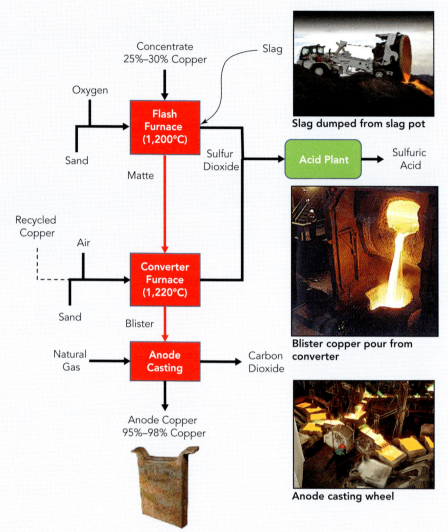

Slag dumped from slag pot

Blister copper pour from converter

Anode casting wheel

Slag pot photo courtesy of Kress Corporation; other photos courtesy of Asarco

The application of oxygen to any sulfide results in the formation of sulfur dioxide, which is collected and converted to sulfuric acid. Sulfuric acid has several industrial uses, such as in the production of fertilizer and to make coatings in the manufacture of paper. It can also be used to leach low-grade copper ores, a process that will be described later in this chapter.

Slag is utilized to make grit used in grinding and to make building materials such as blocks. Depending mostly on the price of copper, the slag might contain sufficient copper to be worth recovering.

FOR THE CHEMICALLY INCLINED

The chemical reactions that occur in copper smelting vary depending on the amounts of chalcopyrite and other copper and iron sulfides in the concentrate. The following reactions are representative. In the flash furnace, the sulfur in the chalcopyrite is oxidized as follows:

chalcopyrite	+	oxygen	→	iron oxide (solid)	+	sulfur dioxide (gas)	+	iron sulfide (matte)	+	chalcocite (matte)
$CuFeS_2$	+	$3.25O_2$	→	$1.5FeO$	+	$2.5SO_2$	+	$0.5FeS$	+	Cu_2S

The iron oxide reacts with silica to produce slag that floats on top of the matte:

iron oxide	+	silica	→	iron silicate (slag)
FeO	+	SiO_2	→	Fe_2SiO_4

In the converter, the reactions are

iron sulfide (matte)	+	oxygen	+	silica	→	iron silicate (slag)	+	sulfur dioxide (gas)
$2FeS$	+	$3O_2$	+	SiO_2	→	Fe_2SiO_4	+	$2SO_2$

chalcocite (matte)	+	oxygen	→	copper (blister)	+	sulfur dioxide (gas)
Cu_2S	+	O_2	→	$2Cu$	+	SO_2

Sulfur is oxidized from S^{2-} in FeS to S^{4+} in SO_2, and copper is reduced from Cu^+ to Cu in Cu_2S. These reactions are exothermic, meaning they generate heat, which aids the smelting process.

Refining

Anode copper is only 95%–98% pure copper, which is not pure enough for applications. To make it pure enough, it must be refined, and electrolysis is used to do this. The cast anode is placed in a tank and connected to the positive side of a power supply (Figure 3.22). The negative side of the power supply is connected to a thin starter plate of pure copper. The tank is filled with a solution containing copper sulfate and sulfuric acid. This is a conductive solution called an *electrolyte* and it completes the electrical circuit.

FIGURE 3.22

Electrorefining of copper anode

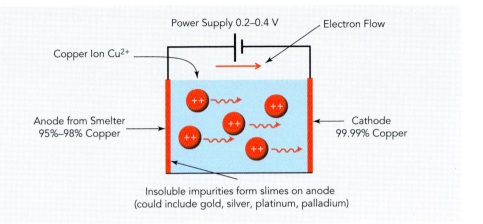

Application of a voltage (less than 1 V, or volt) causes electrons to leave the copper atoms in the anode and travel through the power supply to the cathode. The copper ions (Cu^{2+}) migrate through the electrolyte and plate onto the cathode; that is, the copper ions receive two electrons at the cathode to become a copper atom.

After 10 to 14 days in the cell, enough copper has accumulated on the cathode to form a cathode copper plate, which is sold to buyers such as wire, rod, and pipe manufacturers. The cathodes are "harvested" from the cell (Figure 3.23).

FIGURE 3.23

Cathode copper being harvested from refinery cell

Courtesy of Asarco

Smelting and Refining of Other Metal Concentrates

The smelting and refining process is similar for other metal sulfide concentrates. The metal is produced by smelting the concentrate (oxidizing the sulfide) and has to be refined. In each case, sulfur dioxide is produced and converted to sulfuric acid.

Lead concentrate is oxidized in a flash smelter at about 1,000°C to produce lead oxide (PbO) and sulfur dioxide. The lead oxide is then mixed with coke (carbon) to produce liquid lead bullion and carbon dioxide. The lead bullion may contain many other metals such as copper, arsenic, antimony, bismuth, silver, and gold. The copper occurs as chalcopyrite, which forms a matte on top of the lead once it is cooled to 400°C. The other metals are recovered as anode slimes formed during the refining of the lead bullion.

Some zinc concentrate is roasted to produce zinc oxide, which is treated with sulfuric acid (H_2SO_4) to produce a zinc sulfate ($ZnSO_4$) solution and a residue that may contain metals used in the semiconductor industry such as indium and germanium. Most of the zinc concentrate is pressure leached in an autoclave to produce a zinc sulfate solution. The zinc sulfate solutions produced contain zinc as the ions Zn^{2+}, which can be forced to deposit on a cathode by application of an electric current to the solution. This is called *electrowinning*, which is similar to electrorefining except that the metal is "won" from the solution. (Pressure leaching and electrowinning are described in the "Hydrometallurgy" section that follows.)

Lead-zinc concentrates contain many other valuable metals besides lead and zinc. For example, the smelter at Trail in southeastern British Columbia (Canada) receives lead and zinc concentrates from the Red Dog mine in Alaska (United States) and the Antamina mine in Peru. From these concentrates, the smelter produces refined lead, zinc, silver, and gold as well as a variety of specialty metals such as germanium (used in semiconductors), indium (used in flat-screen TVs), cadmium (used in rechargeable batteries), bismuth, and antimony.

Nickel sulfide deposits contain the mineral pentlandite [$(Ni,Fe)_9S_8$] and other sulfide minerals such as pyrrhotite (an iron sulfide), and chalcopyrite. Cobalt and the platinum group metals, platinum, palladium, and ruthenium, also occur in small amounts. A nickel sulfide concentrate is oxidized in a flash furnace, resulting in a nickel sulfide matte that contains iron oxide. The iron oxide is converted to slag in a converter furnace.

Several processes are used to obtain pure nickel from the nickel sulfide matte. Roasting in a fluidized bed converts the nickel sulfide to a nickel oxide and finally to nickel. A second process uses chlorine to obtain a solution of nickel chloride that can be used to obtain pure nickel by electrowinning. A third process uses ammonia and oxygen to obtain a solution of nickel ammonium sulfate from which nickel is precipitated by the addition of hydrogen gas. These processes are used in different parts of the world and each has advantages, particularly in its ability to recover the different metals in the concentrate.

HYDROMETALLURGY

Hydrometallurgy (or "hydromet" for short) is defined as:

> The use of aqueous solutions, oxygen, and other substances to recover metals from ores, concentrates, and recycled or waste materials

The application of the aqueous solutions is done in heap leach pads, vats, or autoclaves (pressure vessels).

Hydrometallurgy can be applied to the recovery of almost any metal from a variety of sources, including metal and electronic scrap and tailings. In many ways, hydromet will be the future of the metal extraction industry. The wide range of applications makes hydromet an enormous topic. The focus in this chapter is on applications to copper and gold.

Leaching Reactions

The first step in the hydrometallurgical treatment of ore is called *leaching*. This is where the aqueous reagent is applied to the ore rock to ionize the metal and put it into solution. Leaching can be applied to ore stacked in heaps, in vats, or under pressure in an autoclave. Any mineral can be decomposed by using an appropriate reagent, but the most common applications are to copper and gold ores, which are described here in some detail.

Sulfuric acid (H_2SO_4) is applied to low-grade copper sulfides and oxides (from the secondary sulfide zone in a copper deposit). As illustrated in the following graphic,

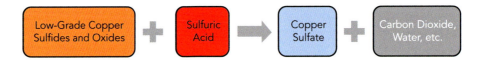

this results in a solution of copper sulfate [$CuSO_4$ ionized as Cu^{2+} and $(SO_4)^{2-}$] and a number of different compounds, depending on the particular sulfide or oxide. Some of these reactions proceed slowly or use large amounts of acid.

When a solution containing the cyanide ion CN^- is applied to gold ore, it forms a stable water-soluble complex with gold, an aurocyanide ion. (Any silver in the ore forms a silver cyanide complex.) For gold, the chemical reaction is called the *Elsner reaction*. It results in a solution containing the aurocyanide complex and sodium hydroxide:

gold	+	sodium cyanide	+	water	+	oxygen	→	sodium aurocyanide	+	sodium hydroxide
4Au	+	8NaCN	+	$2H_2O$	+	O_2	→	$4NaAu(CN)_2$	+	4NaOH

Leach Pads

For low-grade ores, leaching is commonly done in leach pads, which are mounds of ore constructed in *lifts*, each about 5–15 m in height. An example is shown in Figure 3.24. In situ leaching can be done—that is, applied directly to the

FIGURE 3.24

Conventional flat leach
pad in northern Chile

Source: Thiel and Smith 2003, with permission from Thiel Engineering

unexcavated ore—but this can only be applied under certain conditions and it is difficult to control.

On a leach pad, the leaching is done by pumping a dilute solution of the reagent, called a *lixiviant*, through a drip trickle irrigation system placed on the upper lift. The solution percolates through the heap, producing a *leachate* (also known as *pregnant leach solution*) that exits at the base of the pad. Placement of a new lift is done after the concentration of the leachate drops below a particular value. The time between lift placements could be several months.

Copper Hydrometallurgy

Solvent extraction and electrowinning, or SX/EW, is a method for concentrating copper in the leachate that results from using sulfuric acid to leach low-grade secondary copper sulfides and oxides. Flotation, smelting, and refining of these low-grade ores would not be economic, given the low grade of the ore, nor possible because oxide ore cannot be floated.

The first stage of SX/EW is solvent extraction (SX), which involves redissolving the copper from the leachate using an organic solvent and then stripping the copper from the solvent with a high-concentration solution of sulfuric acid. The result is a high-concentration pure solution of copper sulfate. The second stage of SX/EW is electrowinning (EW) in which the copper in the solution can be extracted by electrolysis.

The SX stage is illustrated in Figure 3.25. The leachate, which contains about 1 g (gram) of copper per liter (Cu/L), is pumped into a mixing tank where it is mixed with an organic solvent, an acid labeled HR to denote a hydrogen atom (H) and a long-chain hydrocarbon (R). (The solvent is lighter than water and floats.) The copper sulfate in the leachate and solvent react in the mixer as follows:

copper sulfate	+	organic acid	$\rightarrow$	loaded organic	+	regenerated sulfuric acid
$CuSO_4$	+	$2HR$	$\rightarrow$	CuR_2	+	H_2SO_4

FIGURE 3.25

Solvent extraction process for concentrating copper leachate

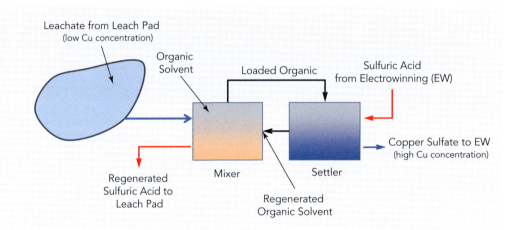

Leachate from Leach Pad (low Cu concentration)

Organic Solvent

Loaded Organic

Sulfuric Acid from Electrowinning (EW)

Copper Sulfate to EW (high Cu concentration)

Regenerated Sulfuric Acid to Leach Pad

Mixer

Settler

Regenerated Organic Solvent

The regenerated sulfuric acid is returned to the heap leach pad. The copper organic phase, which contains about 5 g Cu/L in solution, goes to another tank called a *settler* where it is mixed with a stronger acid solution to strip the copper from the CuR_2 as follows:

loaded organic	+	sulfuric acid	$\rightarrow$	copper sulfate	+	regenerated organic acid
CuR_2	+	H_2SO_4	$\rightarrow$	$CuSO_4$	+	$2HR$

Now the copper sulfate solution is much richer in copper, 30–50 g Cu/L. The organic acid is recovered and reused.

The rich copper sulfate solution is transferred to a tank in which electrolysis is used to "win" the copper from the solution, as shown in Figure 3.26. This looks very similar to electrorefining, but the difference is that the copper is in solution as copper ions, Cu^{2+}, not in the anode as a solid. A voltage (about 2 V) is applied to break down water molecules at the anode, releasing electrons (an oxidation of hydrogen):

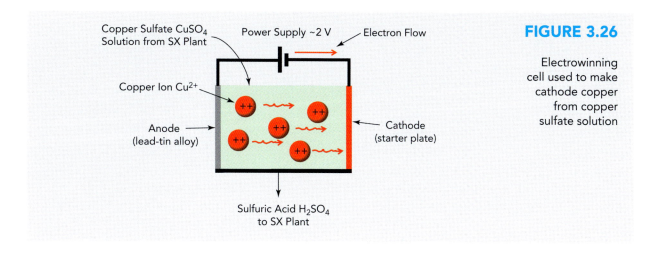

FIGURE 3.26

Electrowinning
cell used to make
cathode copper
from copper
sulfate solution

water	→	hydrogen ions	+	oxygen	+	two electrons
H_2O	→	$2H+$	+	$0.5O_2$	+	$2e$

Being positively charged, the copper ions are attracted to the negatively charged cathode where they combine with the electrons to form copper metal (reduction of copper ions):

copper ion	+	two electrons	→	copper metal
Cu^{2+}	+	$2e$	→	Cu

The result is cathode copper that is as much as 99.999% (five nines) pure.

Oxygen is formed at the anode and produces bubbles. In addition, the hydrogen ions, H^+, combine with the sulfate ion, SO_4^{2-}, to produce sulfuric acid, H_2SO_4, in the tank. When the bubbles reach the surface, they burst, liberating an aerosol of sulfuric acid, an acid mist. This is not good for the health of operators in the tank house. Chemical additives are used to reduce the size of the bubbles, and a thin layer of foam is placed over the electrolyte to keep the bubbles from reaching the surface.

EW requires much more energy than electrorefining because more energy is required to oxidize water and provide two electrons than to oxidize copper to the Cu^{2+} state from a copper anode and provide two electrons—copper gives up electrons easily, whereas water does not. Research is being done to reduce the energy

THE HISTORY OF SOLVENT EXTRACTION AND ELECTROWINNING

The origins of the solvent extraction/electrowinning (SX/EW) process date back to 1915 when copper was leached from ore using acid at Ajo, Arizona (United States), and at the Chuquicamata mine in Chile. Even though the concentration of copper in solutions was low, electricity was cheap at the time and EW was used to obtain copper cathode. However, the result was not pure enough and had to be smelted and refined.

The problem was trying to obtain a high-concentration copper sulfate solution without the other metals that might be present in the ore. The required innovation came from outside the mining industry. During the 1950s, the chemical division of General Mills, a food company, was involved in research into the development of organic chemicals called *chelators* that formed highly selective bonds with metals. (The word *chelator* is derived from the Greek word for "claw.") At that time, the chemical division successfully developed a chelator used in the recovery of uranium.

By 1962, the division had developed an organic solvent called *LIX*, capable of producing pure, high-concentration solutions of copper sulfate. In 1968, a company called *Ranchers Exploration* used LIX in the first commercial SX/EW plant built at its Bluebird mine near Miami, Arizona (United States). In 1970, the Bagdad mine built a second plant, which is still in operation.

In 1964, the director of research and development of a large copper producer predicted that there would never be a pound of copper recovered using solvent extraction. His comment prompted applause. In 2022, about 18% of the world's copper supply was produced using SX/EW.

Source: Bartos 2002; Kordosky 2002

requirement of EW by changing the anode reaction to one that provides electrons more easily than oxidizing water.

The SX/EW facility at the Bagdad mine in Arizona is shown in Figure 3.27a. The pregnant leach solution is derived by leaching low-grade ore piles as well as by leaching waste dumps. Cathodes formed after 5 days in the EW tank are "harvested," as shown in Figure 3.27b.

Pressure Leaching of Sulfide Concentrates

Pressure leaching can be used to treat sulfide concentrates, converting them to metal sulfate solutions from which the pure metal is obtained by EW. The advantage of pressure leaching is that it can be done at lower temperatures, although high pressures are required. The leaching is done in an autoclave (a cylindrical steel pressure vessel) where concentrate slurry is agitated or stirred for about 30 minutes. The temperatures used range between 100° and 232°C, and the pressures range between 1,379 and 4,137 kPa. For chalcopyrite and molybdenite, the process is illustrated in Figure 3.28.

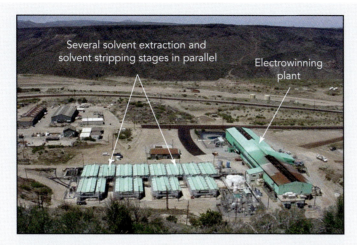

FIGURE 3.27

Solvent extraction/ electrowinning (SX/EW) facility at the Bagdad mine, Arizona

(a) SX/EW plant

(b) Cathode harvesting in the EW plant

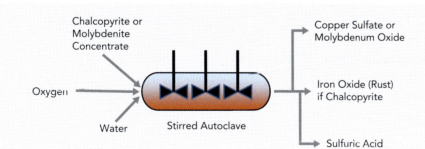

FIGURE 3.28

Pressure leach process for chalcopyrite and molybdenite concentrates

For chalcopyrite concentrate, two chemical reactions occur:

chalcopyrite	+	oxygen	→	copper sulfate	+	ferrous sulfate
$CuFeS_2$		$4O_2$	→	$CuSO_4$	+	$FeSO_4$

ferrous sulfate	+	oxygen	+	water	→	ferric oxide (rust)	+	sulfuric acid
$4FeSO_4$	+	O_2	+	$4H_2O$	→	$2Fe_2O_3$	+	$4H_2SO_4$

The copper sulfate is sent to an EW plant to make cathode copper. Impurities, such as antimony, bismuth, arsenic, and mercury, are found within the iron oxide (rust) that precipitates during the leach. Precious metals in the concentrate can also be found in the iron oxide and can be extracted using a cyanide leach process.

For molybdenite concentrate, the chemical reaction is

molybdenite	+	oxygen	+	water	→	molybdenum trioxide	+	sulfuric acid
MoS_2	+	$4.5O_2$	+	$2H_2O$	→	MoO_3	+	$2H_2SO_4$

The molybdenum trioxide is treated with hydrogen to form water and pure molybdenum, which is used to harden steel and in corrosion-resistant alloys and has applications in electrochemical and display devices. (In contrast, molybdenum sulfide concentrate is roasted to produce molybdenum trioxide and sulfur dioxide.)

Zinc sulfide concentrates are pressure leached using oxygen and sulfuric acid to produce solid sulfur and a solution of zinc sulfate from which pure zinc can be produced by EW. Lead sulfide concentrates, which are usually processed in the same facility as zinc sulfide concentrates (e.g., the Trail smelter in British Columbia), are smelted, resulting in lead bullion and sulfur dioxide. The sulfur dioxide is used to make sulfuric acid, which is used for pressure leaching zinc sulfide concentrates.

The sulfuric acid from all of the preceding reactions can also be used in other leaching processes (heap leaching of copper ores or pressure leaching of zinc sulfides), fertilizer production, and many other industrial and chemical manufacturing processes.

Gold Ore Processing

Gold is leached with a sodium cyanide solution (about 0.5 g/L), either in heap leach pads or in tanks, and via the Elsner reaction. This produces a leachate solution of sodium aurocyanide, which contains the gold ion Au^+ in solution. A similar reaction occurs with silver. There are two ways to obtain gold from this solution. In one method, zinc is used to precipitate the gold, whereas in the other, the aurocyanide is adsorbed onto activated carbon.

If leaching conditions are acidic to slightly alkaline, the following chemical reaction will occur:

sodium cyanide	+	water	→	hydrogen cyanide	+	sodium hydroxide
$NaCN$	+	H_2O	→	HCN	+	$NaOH$

Hydrogen cyanide is a poisonous gas. Its release into the atmosphere also results in a loss of cyanide, which is an expensive reagent. For these reasons, lime is added to the ore to make the leaching conditions highly alkaline and prevent the above reaction.

Merrill–Crowe Process

The Merrill–Crowe process for producing gold (or silver) from the aurocyanide solution precipitates gold by the addition of zinc dust. The entire process is illustrated in Figure 3.29. High-grade ore (>5 g/t) passes through a crushing and grinding stage before being pumped into a series of leach tanks containing cyanide. Low-grade ore (1–2 g/t) is stacked onto a leach pad and leached with a dilute solution of cyanide and lime distributed through a drip trickle irrigation system.

The pregnant solution from the heap leach pad is often full of suspended solids that interfere with the process. These are removed using a clarifier (described in a later section) and a screen.

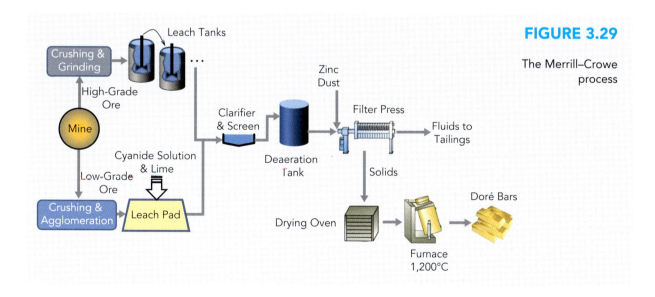

FIGURE 3.29

The Merrill–Crowe process

The precipitation of gold and silver from the solution is done using zinc. The process is known as *zinc cementation* and consists of two chemical reactions (shown here for the case of gold):

		zinc	+	sodium aurocyanide	$\rightarrow$	gold	+	sodium zinc cyanide		
		Zn	+	$2NaAu(CN)_2$	$\rightarrow$	$2Au$	+	$Na_2Zn(CN)_4$		

zinc	+	sodium cyanide	+	oxygen	+	water	$\rightarrow$	sodium zinc cyanide	+	sodium hydroxide
Zn	+	$4NaCN$	+	$\frac{1}{2}O_2$	+	H_2O	$\rightarrow$	$Na_2Zn(CN)_4$	+	$2NaOH$

Zinc dust is added and the resulting solids are filtered, producing a barren waste solution, and then smelted to produce a doré bar weighing about 25 kg. Oxygen is removed from the aurocyanide solution (deaeration) to stop the second reaction that consumes zinc. The amount of gold in a doré bar varies widely and depends on the composition of the ore body.

The Merrill–Crowe process is used when the silver-to-gold ratio of the ore is high because silver cannot be recovered using adsorption onto activated carbon, a process described in the following section. However, if the ore contains a large amount of clay, the filtering process can become difficult.

WHY ZINC?

Zinc gives up electrons (oxidizes) much more readily than gold or silver (see Table A.1 in Appendix A).

A gold or silver ion in solution will pick up any electrons zinc provides and precipitate to form the solids that are used to make the doré bar. Zinc is also readily available and relatively inexpensive.

Adsorption of Aurocyanide onto Activated Carbon

Activated carbon is produced by charring materials, such as wood, coconut shells, or peach pits, at temperatures between 700° and 800°C. Steam and/or chemicals are used to develop microporosity. The result is enormous internal surface areas where adsorption can occur; 1 g of activated carbon has 500 m^2 of surface area.

Activated carbon is used in water filters to adsorb impurities, but it can also adsorb the negatively charged aurocyanide ion $[Au(CN)_2]^-$ onto the surfaces of the micropores in the carbon. This is used to concentrate the aurocyanide ion in the leachate solution. The result is loaded carbon with gold concentrations up to 200,000 g/t.

The aurocyanide is stripped from the carbon, resulting in a high concentration of gold in solution from which solid gold can be obtained by EW. The carbon is regenerated and reused by heating it in a kiln.

There are three leaching and adsorption processes by which gold can be extracted. These are illustrated in Figure 3.30. The method used for a particular gold ore depends almost entirely on the ore grade and the other minerals in the ore.

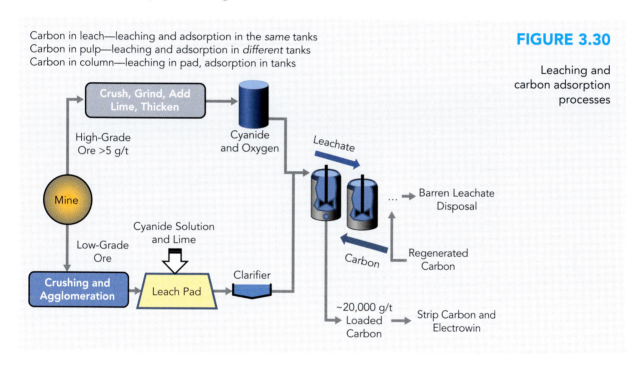

Carbon in leach—leaching and adsorption in the *same* tanks
Carbon in pulp—leaching and adsorption in *different* tanks
Carbon in column—leaching in pad, adsorption in tanks

FIGURE 3.30

Leaching and carbon adsorption processes

If the grades in the mine are high (>5 g/t), the ore is ground into fine particles (typically 75 μm or less); mixed with lime, cyanide, and oxygen; thickened; and then transferred into mechanically stirred tanks where leaching and adsorption occur. In the *carbon-in-leach* (CIL) process, leaching and adsorption occur in the same tanks, whereas in the *carbon-in-pulp* (CIP) process, leaching and adsorption occur in separate tanks.

Note in Figure 3.30 that the activated carbon flows in the opposite direction to the leachate slurry flow—a *countercurrent flow*—so that fresh carbon is present when the gold concentration in the leachate is low, allowing maximum recovery in each adsorption tank. This is a common feature of all carbon adsorption processes.

ADSORPTION VERSUS ABSORPTION

Adsorption is a surface phenomenon in which molecules are concentrated on the surface of a substance, whereas *absorption* is what a towel or sponge does to water. Any two substances brought close together will experience an attraction as the negative electrons of one are attracted to the positive nuclei of the other. Aurocyanide is negatively charged and is attracted to the positive nuclei of the carbon atoms. The attractive force is additive, meaning that the greater the availability of carbon atoms, the greater the attraction. Thus, attraction depends on available surface area.

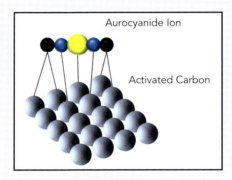

The CIP process can achieve gold recovery of more than 95%. However, the increase in recovery must be balanced with the higher capital costs of two sets of tanks. (The tanks are lined with stainless steel.) Also, if there is naturally occurring carbon in the ore (a quite common occurrence), it will compete with the activated carbon; in addition, any silver or copper minerals present will also adsorb gold. All of these phenomena are known as *preg-robbing* because they take the aurocyanide in the pregnant (value-bearing) solution from the activated carbon "without asking permission."

The advantage of CIL is lower capital costs. CIL is used when the ore is naturally carbonaceous to force adsorption onto the activated carbon. However, leaching and adsorption in the same tank leads to concentration gradients that must be broken down. This is done using greater agitation than that required in CIP tanks. The result is loss of precious metals from the carbon and lower recovery than in CIP.

If the gold grades in the mine are low (1–2 g/t), the ore is first crushed and may be agglomerated, after which it is stacked onto a heap leach pad and leached with a dilute solution of cyanide and lime distributed through a drip trickle irrigation system. The pregnant leach solution, containing 0.5–5 mg/L gold, is clarified and pumped through mechanically stirred tanks where carbon adsorption of the auro-cyanide occurs. This is known as the *carbon-in-column* (CIC) method. Five tanks of a group of six CIC tanks are shown in Figure 3.31.

FIGURE 3.31

CIC tanks. Looking in the direction of leachate flow. Activated carbon flows in the opposite direction and adsorbs gold in the leachate.

Courtesy of Morrison-Maierle

Leaching and carbon adsorption are the concentration processes that put the gold into solution and produce loaded carbon from which the gold is to be obtained. Figure 3.32 is a schematic of the carbon stripping and EW process that occurs after carbon adsorption. First, the aurocyanide is stripped from the carbon using some combination of water, caustic soda, and acid. Next, the stripped or regenerated carbon particles are dried in a rotary kiln and recycled to the tanks. The aurocyanide solution, containing 50–1,000 mg/L gold, is then pumped to an electrolytic cell where the gold is won from the solution by applying an electric current.

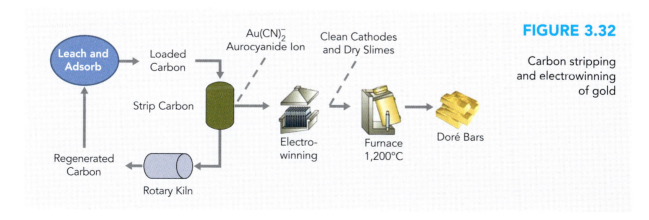

FIGURE 3.32

Carbon stripping and electrowinning of gold

At the anode or positive electrode, oxygen is produced by the breakdown of water, yielding some electrons:

water	→	hydrogen ion	+	oxygen	+	four electrons
$2H_2O$	→	$4H^+$	+	O_2	+	$4e$

At the cathode or negative electrode, deposition of gold occurs at voltages between 0.7 and 1.1 V.

aurocyanide ion	+	electron	→	gold	+	cyanide ion
$Au(CN)_2^-$	+	e	→	Au	+	$2CN^-$

After as much as 1 hour in the cell, the cathodes are removed and washed (Figure 3.33). The solution might be recirculated several times to improve gold recovery. Steel wool may also be used as a cathode.

FIGURE 3.33

Washing impure gold off cathode (yes, that is a garden hose)

The gold (or steel wool and gold) is then smelted with fluxes, such as silica and borax, which remove base metal oxides, and then poured into doré ingots that contain varying quantities of silver (Figure 3.34). The ingots are sent to a refinery to be made into pure gold.

Alternatives to Cyanide

Owing to the hazards associated with cyanide, there has been considerable research into the possible use of alternatives for leaching gold. Most of the research has

FIGURE 3.34

Pouring doré bars
at a gold mine

Source: Agnico-Eagle Mines Limited 2006, CC0 1.0

focused on the use of thiosulfate [ionic formula $(S_2O_3)^{2-}$], thiourea (chemical formula NH_2CSNH_2), and halides such as chlorine. Of these three, thiosulfate appears to be the most promising. Unfortunately, the thiourea reagent doses required are high and thiourea is a potential carcinogen. Gold halides also break down rather easily, and considerable process control is required for halide leaching. The major obstacle to these alternatives is that no economic method for winning the gold from solution has been found.

Biohydrometallurgy

Biohydrometallurgy is the use of bacteria to facilitate the breakdown of sulfide minerals. The breakdown is a form of oxidation, and thus the term *biooxidation* arises. Some of the products of the oxidation (e.g., metals) may become dissolved, and this gives rise to the term *bioleaching*. The distinction is whether a product goes into solution or not. The term *biomining* is more general and refers to both processes.

Bacteria can also be used to change solution chemistry to recover metals or treat metal-contaminated water. The focus in this section is on processing metal sulfides.

This is a very brief summary of a complex but interesting subject. Bacteria are able to survive hostile environments, medium to high temperatures, and high acidity. Over the eons, they have evolved mechanisms to deal with these conditions, survive, and play a significant role in the formation of mineral deposits and in the earth's sulfur cycle, part of which is sulfide dissolution.

Sulfide dissolution occurs naturally but slowly. For example, a common natural oxidation process is the breakdown of pyrite in the presence of oxygen and water:

pyrite	+	oxygen	+	water	→	ferrous iron	+	sulfate ion	+	hydrogen ions
$2FeS_2$	+	$7O_2$	+	$2H_2O$	→	$2Fe^{2+}$	+	$4SO_4^{2-}$	+	$4H^+$

This is an oxidation of a sulfide to the sulfate ion, SO_4^{2-}. The reaction proceeds slowly, but certain bacteria derive energy from the oxidation and will catalyze the reaction. The same bacteria as well as other bacteria derive energy from the oxidation of the ferrous iron to ferric iron in the presence of oxygen:

ferrous iron	+	oxygen	+	hydrogen ions	→	ferrous iron	+	water
$4Fe^{2+}$	+	O_2	+	$4H^+$	→	$4Fe^{3+}$	+	$2H_2O$

Ferric iron is a strong oxidizer and causes further oxidation of the pyrite. Thus, directly and indirectly, bacteria accelerate the breakdown of pyrite by as much as half a million to a million times.

There are several bacterial strains that oxidize ferrous iron, and there are others that oxidize sulfur compounds resulting from the oxidation of a metal sulfide. All of them may be present in a biomining operation. All have names that are difficult to pronounce; therefore, their names are abbreviated. Some examples:

- *Acidithiobacillus ferrooxidans* (Af): iron or sulfur oxidizer
- *Acidithiobacillus thiooxidans* (At): sulfur-only oxidizer
- *Acidithiobacillus caldus* (Ac): sulfur-only oxidizer
- *Leptospirillum ferrooxidans* (Lf): iron-only oxidizer

Like every living thing, these bacteria need carbon and nitrogen to grow sugars and other molecules to form new cell mass. Carbon can be obtained from carbon dioxide in the air used to aerate the ore slurry or by means of addition of a carbon source such as yeast. Nitrogen can be obtained from the air or by addition of fertilizer to slurry. Bacteria also need energy to grow, which they derive by oxidizing the ferrous ion to ferric ion in the presence of oxygen or by oxidizing inorganic sulfur compounds. Considerable heat is generated by the oxidation process and an acidic solution full of metal ions is created. To different degrees, all the preceding strains have evolved intricate biochemical systems to obtain carbon, oxidize iron or sulfur, and to tolerate the heat, acidity, and metal ion concentrations.

The oxidation of the metal sulfide, MS, puts the metal into solution as the ion M^{2+}. Sulfur compounds, such as thiosulfate ($S_2O_3^{2-}$), and pure sulfur also result. These are oxidized by the sulfur-oxidizing bacteria present. Some sulfides are soluble in acid and the hydrogen ions generated during oxidation contribute to further dissolution of the sulfide.

Where does all this chemistry occur? The bacteria form an exopolysaccharide (EPS) layer around themselves when they come into contact with a metal sulfide. Figure 3.35 shows the configuration for leaching of pyrite. It is within the EPS layer, not in the solution, that the oxidation and leaching reactions take place. The bacteria have evolved to withstand the high acidity and high metal content of the solution.

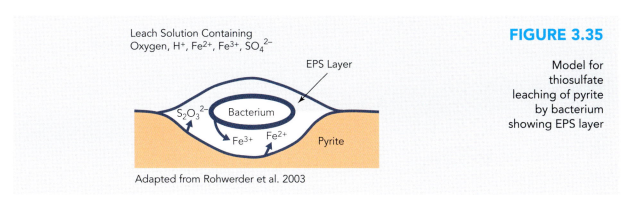

Leach Solution Containing Oxygen, H^+, Fe^{2+}, Fe^{3+}, SO_4^{2-}

EPS Layer

$S_2O_3^{2-}$ Bacterium

Fe^{3+} Fe^{2+} Pyrite

Adapted from Rohwerder et al. 2003

FIGURE 3.35

Model for thiosulfate leaching of pyrite by bacterium showing EPS layer

Biomining is done in a series of stirred, aerated tanks or in a heap leach pad. Stirred tank systems operate continuously; the output from one tank is the input for the next. An example is shown in Figure 3.36. The tanks are expensive—made of stainless steel or a ceramic composite to avoid corrosion. Consequently, only high-grade, high-value ore is processed this way.

A common application is refractory gold ore where the gold is contained within sulfides such as arsenopyrite. Biooxidation of the sulfides releases the gold as particles in the concentrate. Fine particles of a sulfide concentrate resulting from sulfide flotation are added to the first tank. The slurry in each tank is stirred to ensure that the particles remain in suspension. The tanks are also aerated to promote growth of the bacteria as well as to maintain the temperature within limits that the bacteria can tolerate (30°–45°C). Nutrients, such as phosphate and ammonium, are added to the tanks to promote bacterial growth. In the final tank, the concentrate is thickened and then sent to the cyanidization process.

Bio–heap leaching is typically applied to low-grade copper ores. It can accelerate the leaching reactions of the SX/EW process so that the leach time of a lift on a heap is reduced to a few months rather than almost a year or more. Leaching bacteria are

FIGURE 3.36

Stirred bioreactors at a copper sulfide leach project

Courtesy of BacTech Environmental

present naturally, and it is common to add fertilizers to the ore to promote growth of these bacteria.

Bio–heap leaching of nickel and cobalt ore is also done. One significant operation is in Kasese, Uganda, where the tailings produced by a closed copper mine are bio-leached. Bioleaching of chalcopyrite is a slow process because layers of sulfur and sulfate minerals develop on the surface of chalcopyrite and inhibit the ability of the bacteria to oxidize the sulfide. At higher temperatures, 60°–75°C, the growth of these layers is inhibited, and different bacteria that can tolerate such temperatures must be used. The bacterial species used for chalcopyrite bioleaching are the subject of a considerable amount of research.

PHYSICAL SEPARATION METHODS

Shaking Table

A shaking table is used to separate any collection of particles with significant differences in mass. Common applications include separation of gold, chromite, and heavy mineral sands that may contain titanium, tin, tungsten, and rare earths.

A shaking table consists of a sloping deck with a riffled surface (Figure 3.37). A motor drives a small arm that shakes the table along its length, parallel to the riffle pattern. The shaking motion consists of a slow forward stroke followed by a rapid return stroke. Water is added at the top of the table perpendicular to the table motion. The riffles cause a barrier to the water flow, which creates eddies, causing

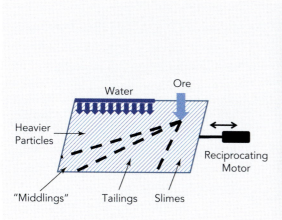

FIGURE 3.37

Shaking table

Courtesy of Overburden Drilling
Management Limited

the water flow to slow down behind a riffle. The heavier particles will fall to the bottom behind a riffle and lighter particles will wash over the riffles with the water flow. Thus, the heaviest and coarsest particles concentrate at the upper end of the table while the lightest and finest particles tend to concentrate at the bottom edge. Intermediate size and density particles or "middlings" are deposited at points in between. Ideally, there should be no heavy particles in the tailings or slimes.

A shaking table is a motorized version of the sluice box used for recovery of gold from river sands (Figure 3.38). The principle is the same—namely, riffles form a barrier to flow, causing heavier gold particles to collect behind them. Normally, most of the trapped gold will be behind the first few riffles. Small flour gold may collect behind several riffles further downstream. The last few riffles should have no gold.

FIGURE 3.38

Sluice box

Courtesy of C. Ralph, Nevada Outback Gems

Centrifugal Concentrator

A centrifugal concentrator is used to recover gold and other precious metals from concentrates. It consists of a riffled cone or bowl that spins at high speed to create forces in excess of 60 times that of gravity (Figure 3.39). A slurry mixture is introduced into the cone through a central tube; the centrifugal force produced by rotation drives the solids toward the walls of the cone. The slurry migrates up along the wall where heavier particles are captured within the riffles. Lighter particles rise to the top of the cone and over its side. Injecting water from a cavity through holes located in the back of the riffles fluidizes the riffled area preventing compaction of the concentrate bed and forcing the heavy minerals to the bottom of the bowl.

FIGURE 3.39

Cross section of a centrifugal concentrator

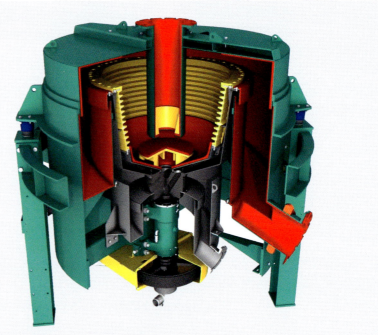

Courtesy of FLSmidth, © FLSmidth A/S

Spiral Concentrator

A spiral concentrator is a tower around which is wound a sluice (Figure 3.40). It is used to separate heavier from lighter particles by means of their density differences, but the hydrodynamics of the particle slurry also play a role.

The particle slurry is introduced at the top of the spiral. Heavier particles sink to the bottom of the sluice where they experience more drag and travel slower, causing them to move toward the center of the spiral. Lighter particles stay toward the

FIGURE 3.40

Spiral concentrator

outside of the spiral and remain suspended in the water and, consequently, arrive at the bottom of the spiral faster. The heavier particles come out of suspension and are extracted through slots or channels placed in the base of the sluice. At the bottom of the sluice, a set of adjustable bars, channels, or slots is used to separate the low- and high-density parts.

Spiral concentrators are used to separate dense minerals from sands, fine coal particles, and heavy particles in iron ore. They have no moving parts and are typically made of urethane, which reduces wear. A typical processing rate is between 1 and 3 t/h. Consequently, large numbers of concentrators are used in "spiral farms."

Magnetic Separation

Magnetic fields are used to separate any magnetic mineral such as iron ore from waste nonmagnetic minerals. The ore feed is forced to flow over a hollow cylinder in which a rare-earth permanent magnet is placed (Figure 3.41a). As the drum rotates, the magnetic material is attracted to the magnet and tends to stay attached to it. Toward the bottom of the drum, gravity takes over and the magnetic material falls into a collection area. Arrays of magnetic concentrators may be used, as shown in Figure 3.41b, to allow multiple stages of magnetic separation.

Thickeners and Clarifiers

Thickeners and clarifiers are used to separate liquids and solids by particle settling. Thickeners and clarifiers look alike, but there are differences in their design

FIGURE 3.41

Magnetic separator

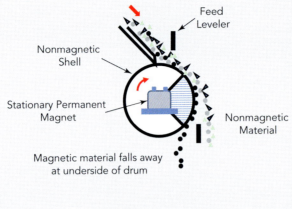

(a) Cross section

(b) Array of concentrators in an iron ore plant (Photo © Metso Corporation)

and operation. The purpose of a thickener is to increase the solids content of feed slurry (from 2%–10% solids to 20%–50% solids), whereas a clarifier removes a small amount of suspended particles to produce clear(er) water.

A thickener/clarifier consists of a large tank with a feedwell at the top and a conical bottom leading to a discharge point called the *underflow* (Figure 3.42). A perimeter drain surrounds the tank. Slurry is fed into the feedwell below the water level and settling begins. Once the particles settle to the bottom, rotating rakes pull the solids into the underflow. Clarified water flows to the perimeter drain. A flocculant is often added to the feed slurry. A flocculant is a polymer that attaches itself to the particles to form chains of larger particles that settle faster.

FIGURE 3.42

Basic thickener/clarifier geometry

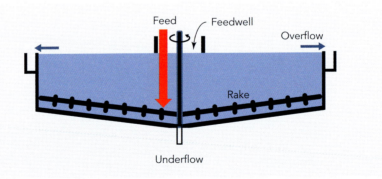

FIGURE 3.43

Tailings thickeners,
Escondida mine, Chile

Courtesy of FLSmidth, © FLSmidth A/S

The degree of thickening is primarily controlled by the surface area or diameter of the thickener. Conventional thickeners that remove only some water are very-large-diameter (>50 m), shallow (10 m) structures (Figure 3.43). Small-diameter, deep cone thickeners having steep conical sides force settlement within a small width, resulting in very high solids density in the underflow. Application of deep cone thickeners is discussed in Chapter 5.

REFERENCES

Agnico-Eagle Mines Limited. 2006. LaRonde gold pour. http://commons.wikimedia.org/wiki/File:LaRonde_Gold_Pour.jpg. Accessed November 2023.

Bartos, P.J. 2002. SX-EW copper and the technology cycle. *Resources Policy* 28:85–94.

Cleary, P.W., and Morrison, R.D. 2012. Prediction of 3D slurry flow within the grinding chamber and discharge from a pilot scale SAG mill. *Minerals Engineering* 39:184–195.

Flores, G.A., Risopatron, C., and Pease, J. 2020. Processing of complex materials in the copper industry: Challenges and opportunities ahead. *JOM* 72(10):3447–3461.

Gupta, A., and Yan, D. 2016. *Mineral Processing Design and Operations: An Introduction*, 2nd ed. Amsterdam: Elsevier. p. 698.

Kordosky, G.A. 2002. Copper recovery using leach/solvent extraction/electrowinning technology: Forty years of innovation, 2.2 million tonnes of copper annually. *Journal of the Southern South African Institute of Mining and Metallurgy* 102(8):445–450.

Monteiro, L.V.S, Xavier, R.P., Hitzman, M.W., et al. 2008. Mineral chemistry of ore and hydrothermal alteration at the Sossego iron oxide–copper–gold deposit, Carajás Mineral Province, Brazil. *Ore Geology Reviews* 34(3):317–336.

Rohwerder, T., Gehrke, T., Kinzler, K., et al. 2003. Bioleaching review: Part A. *Applied Microbiology and Biotechnology* 63:239–248.

Thiel, R., and Smith, M.E. 2003. State of the practice review of heap leach pad design issues. In *Hot Topics in Geosynthetics—IV, Proceedings of the 17th Annual GRI Conference*. Folsom, PA: Geosynthetics Institute.

Workman, L., and Eloranta, J. 2003. The effects of blasting on crushing and grinding efficiency and energy consumption. In *Proceedings of the 29th Conference on Explosives and Blasting Techniques*. Nashville, TN: International Society of Explosives Engineers, pp. 1–5.

Nonmetallic Minerals

There are many nonmetallic minerals of economic interest.

Aggregates, such as gravel and sand, are significant components of concrete and cement. Clays have many uses such as paper coatings, liquid absorption, cosmetics (facial masks, fixatives for fragrances), and drilling lubricants. Feldspars are used for making glass and as fillers or extenders in paint and plastic.

Such minerals fall under the general classification of industrial minerals and are typically nonmetallic. Like other minerals, industrial minerals can be formed in any part of the rock cycle. Industrial minerals are processed using mostly physical separation methods.

The field of nonmetallic or industrial minerals is enormous. This chapter focuses on two minerals: coal and diamonds, two very different forms of carbon. Coal is formed in shallow (less than 1 km) sedimentary environments, whereas diamonds are formed by subjecting carbon to very high pressures in the mantle of the earth at depths greater than 150 km.

COAL

How Coal Forms

Coal formation begins in a swamp or bog. An example is shown in Figure 4.1. Deposition of organic debris forms large deposits of peat. Plate tectonics or melting of glaciers leads to rising sea levels and an ocean may cover the peat bog. Layers of sediment containing mostly silicate minerals will deposit on top of the peat, compressing it.

As the peat is compressed by thicker layers of sediments, gases and fluids are squeezed out of it and a layer of coal forms. About a 20:1 volume reduction is required to form coal. The ocean may advance and recede several times over a 75–100 million-year period, giving rise to the possibility of multiple layers of coal and sediments (Figure 4.2a). The photo in Figure 4.2b shows one coal seam formed between two sedimentary rock layers.

Coal rank refers to the amount of carbon in the coal, and increasing rank is associated with escalating pressure caused by the overlying sediments. Figure 4.3 shows the classification of coal according to rank.

FIGURE 4.1

Burns Bog south of Vancouver, British Columbia (Canada)

Courtesy of Burns Bog Conservation Society

FIGURE 4.2

Formation of one coal seam and an example of a single seam

Ocean Level Rises

Ocean Recedes

Swamp or Bog
Peat

Buried Peat

Coal

75–100 Million Years

(a) Formation of a coal deposit by successive deposition of sediments over organic layers

(b) Point Aconi coal seam, Nova Scotia (Canada) (Photo by Michael C. Rygel 2006, CC BY-SA 3.0)

Coal is composed of *macerals*, which are the organic components of coal. The other components are carbonate, sulfide, sulfate, and silicate minerals including clays. The silicates and clays are incombustible; they are referred to as *ash*.

Three common macerals are vitrinite, inertinite, and liptinite, and they are each derived from plant material. *Vitrinite* is one of the primary components of coal and

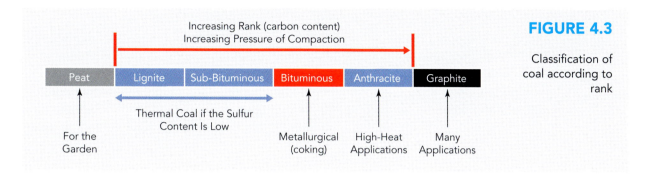

FIGURE 4.3

Classification of coal according to rank

provides the highest heating value per unit mass. It is derived from the cell-wall material or woody tissue of the plants from which the coal deposit was formed. Vitrinite has a shiny appearance and often appears as distinct bands within other macerals (Figure 4.4). *Inertinite* is highly oxidized plant material that may even be burned. *Liptinite* is produced from decayed leaves, spores, and algae, and it retains the original plant form, which makes it easy to recognize.

Courtesy of U.S. Geological Survey

FIGURE 4.4

A specimen of bituminous coal showing the shiny vitrinite macerals

The United States has the largest coal reserves, followed by Russia, China, Australia, and India (EIA 2023). Global trade in thermal and metallurgical coal is reported annually by the International Energy Association (IEA 2023a). From 2021 to 2023, Indonesia exported the largest amount of thermal coal and Mongolia exported the largest amount of metallurgical coal.

Coal Processing

The purpose of processing coal is to remove incombustible material, such as dirt and rock (the ash), to increase the heating value or carbon content of the coal.

Processing coal is sometimes known as *coal washing*. Physical methods, such as screening, dense media separation, flotation (a physico-chemical method), and thickening or drying, are used. A simplified flow sheet for coal processing is shown

IS COAL A MINERAL?

This question can lead to some heated debates. Legally and for tax purposes, coal is considered a mineral. However, strictly speaking, it is not a mineral, even though Skinner (2005) suggests that all solids are potential minerals and gives an expanded definition of a mineral:

> An element or compound, amorphous or crystalline,
> formed through biogeochemical processes

There are biogeochemical processes involved in the formation of coal, but the wide variety of compounds in the original plant remains makes it difficult to define a characteristic chemical composition or set of compounds that make up coal, a feature still required by the above definition. For this reason, geologists refer to coal as a rock—a combination of minerals.

Coal is the official state *mineral* of Kentucky (KGS 2023) and the official state *rock* of Utah (Hylland 1995).

in Figure 4.5. Coal processing begins with *run-of-mine* (ROM) coal from a pit or underground operation. ROM ore consists of coal, rocks, minerals, and sometimes bucket teeth from the shovels used in mining. The moisture content and particle sizes of ROM coal can have a large variability, and this means a coal processing plant must be designed to adapt to changes in these properties. The ROM coal may be crushed to reduce the particle size.

FIGURE 4.5

Flow sheet for a coal processing facility

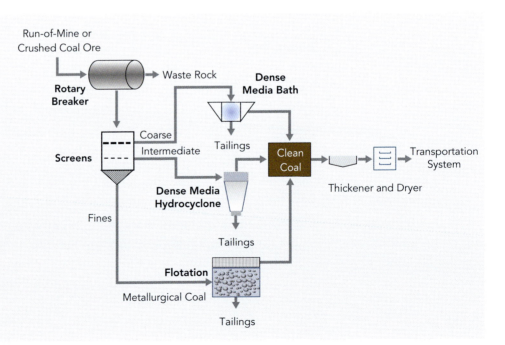

Coal from different parts of the deposit may be taken to the processing plant in particular proportions to achieve a blend of coal that satisfies customer specifications for thermal value as well as ash and sulfur content. The components of the flow sheet are described in the following sections.

Rotary Breaker

A rotary breaker consists of an outer fixed shell and an inner rotating drum with perforations. Typical rotational speed of the drum is 12–18 rpm. A large rotary breaker is shown in Figure 4.6a. Lifter plates inside the breaker (Figure 4.6b) pick up the ROM coal, which then falls onto the drum. The softer coal breaks and passes through the perforations while the harder rock is transported to the waste stream at the discharge end of the breaker. In addition to the cleaning (removal of rock), a size reduction is achieved.

(a) Rotary breaker without dust cover, showing feeder in front

(b) Inside the drum of a rotary breaker showing lifter plates

FIGURE 4.6

A rotary breaker and inside the breaker
(Courtesy of McLanahan Corporation)

Screens

Screens are used to group coal particles into ranges by size. (The size ranges are sometimes called *grades*.) Screens can be static but are commonly mechanically vibrated. The screens are made from different materials such as high-tensile steel, stainless steel, and plastics. Figure 4.7 shows a vibrating screen with two screen decks.

Dense Media Separation

The principle of dense media separation (DMS) is very simple and can be applied to a variety of ores. In the case of coal, a fluid medium, typically magnetite and water, is denser than coal such that when it is mixed with coal slurry, the coal will float to the surface of the fluid and the heavier material will sink. Figure 4.8 shows two types of DMS machines, a drum for separation of coarse particles and a hydrocyclone for finer particles that may not separate easily.

FIGURE 4.7

Vibrating screen with two screen decks

Photo © Metso Corporation

FIGURE 4.8

Dense media separation (DMS) machines (Courtesy of FLSmidth, © FLSmidth A/S)

(a) DMS drum

(b) DMS hydrocyclones

Flotation

Fine coal particles are less than 0.5 mm in size. The fine particles of thermal coal are usually so dirty that they cannot be economically cleaned or separated from ash. They are often discarded, but it might be possible to blend the fines with coarse coal to achieve acceptable ash content. However, the fine particles of metallurgical coal have high carbon content, and they can be separated using flotation to obtain clean coal using different types of oil as a collector. Sometimes the clean fines are agglomerated to form coarse particles.

The total surface area of a volume of fine particles is larger than the surface area of a coarse particle of the same volume. Because heat release from a coal particle is proportional to surface area and there is a large amount of surface area exposed on a collection of fine particles, the fines are desired for both thermal and metallurgical applications. However, during processing and transport, only the surface of the coarse particles oxidizes, whereas an entire fine particle may oxidize, lowering its thermal value. Thus, both thermal and metallurgical coal are ground to fine sizes at the location where they are used.

DIAMONDS

How Diamonds Form

Diamond deposits are found in the oldest parts of continents, called *cratons*, where the basement rocks are older than 1,500 million years. The most productive cratons are older than 2,500 million years, located in the central parts of North America, Asia, India, and Australia. Less productive deposits are found in 1,600–2,500 million-year old rocks.

The location of diamond deposits cannot be related to any plate tectonic activity within the last 100–200 million years. This is because the formation of diamonds and diamond deposits is more related to processes deep in the earth rather than the shallow crustal processes that lead to base and precious metal deposits. In fact, the only way to form a diamond is by subjecting carbon to extremely high temperatures (900°–1,200°C) and pressures that can only be found at depths greater than 150 km in a region below the earth's crust known as the *mantle*.

Once diamonds are formed, they are transported to the surface by magma under considerable pressure (Figure 4.9). Dissolved gases in the magma expand, and the magma combines with boiling groundwater to result in an explosive supersonic eruption at the surface. The high speed prevents the diamonds in the magma from re-crystalizing as graphite. The result is a carrot-shaped pipe or vent at the surface and a small volcanic cone.

The pipes contain dark-colored minerals, such as garnets and pyroxenes, which are formed in the mantle. Fragments of crustal rock are also present. The rock in the pipes is called *kimberlite*, after the city of Kimberley, South Africa, where pipes were first discovered in the 1870s. Pipes occur in clusters and the pipes in a cluster are typically, at most, tens of kilometers apart.

A web search with the terms "kimberlite formation" will result in several videos showing how kimberlite pipes form.

FIGURE 4.9

Diamonds are formed in the earth's mantle and then transported to the crust

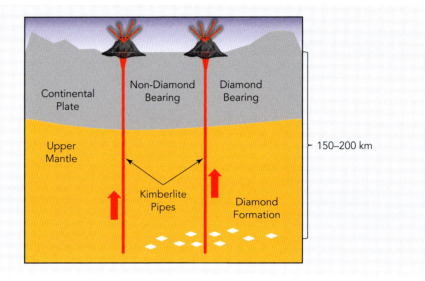

Finding a Diamond Mine in the Arctic

The manner in which diamond deposits were discovered in the Canadian Arctic reveals the kind of lateral thinking geologists do to find ore bodies. Most of northern Canada is underlain by rocks that are 2.5 to 3.5 billion years old; therefore, diamonds should (might) be present in kimberlite pipes. However, the problem is that almost all of Canada was covered by glaciers, which ground up the rock into particles as the glaciers accumulated ice and flowed in a particular direction, and then deposited the particles as *glacial till* when they melted.

HOW OLD ARE DIAMONDS AND WHY ARE THEY RARE?

Diamonds from kimberlite pipes have been age-dated and found to be between 3,300 million and 990 million years old. However, the kimberlite rock was intruded only about 100 million years ago. Given the age of the diamonds, the carbon source is most likely the carbon trapped in Earth's interior at the time Earth formed 4,600 million years ago. The youngest kimberlite pipe in the world is in the Lac de Gras area of Canada and is about 50 million years old.

Two things make diamonds rare: First, only about 1 in 50 kimberlite pipes contain diamonds. Second, explosive eruptions that produce kimberlite pipes seem to have stopped occurring. However, remember that the geological processes that occurred millions of years ago can also occur in the present ("the present is the key to the past"). It is more likely that the cessation of kimberlite pipe formation is only a hiatus (Davis and Kjarsgaard 1997).

Geologist Charles Fipke reasoned that if a kimberlite pipe were present in the rock, the glacier would shear off its top and drag particles of kimberlite as well as any garnets and pyroxenes in the kimberlite pipes along its path of movement. Garnets and pyroxenes are easily recognized, mobile, and resistant to weathering. They are *indicator minerals* that provide strong evidence of the presence of kimberlite. Thus, the presence or absence of indicator minerals in glacial till was the key to finding a kimberlite pipe in the Arctic.

Fipke and another geologist, Stewart Blusson, sampled glacial till at points perpendicular to the path of a former glacier and tracked the trail of indicator minerals, as illustrated in Figure 4.10. The point at which no indicator minerals were found was the likely location of a kimberlite pipe. Plus or minus a few details, some of which become more interesting with time, this was how the first diamond mine discovery, the Ekati mine, in the Northwest Territories (Canada) was made.

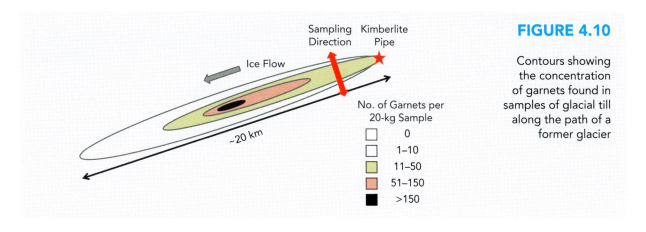

FIGURE 4.10

Contours showing the concentration of garnets found in samples of glacial till along the path of a former glacier

A good book about this diamond find is *Fire into Ice: Charles Fipke and the Great Diamond Hunt* (Frolick 1999). Canada is now one of the largest diamond producers (based on value) in the world.

Diamond Ore Processing

There are two interesting aspects of diamond ore processing. One is how kimberlite ore is crushed, and the other is how the diamonds are separated from the crushed ore.

Comminution of diamond ore must be done carefully, otherwise the diamonds themselves will be crushed or damaged. A high-pressure grinding roll (HPGR) is used to crush the ore. As shown in Figure 4.11, an HPGR machine consists of a pair of counter-rotating rolls, one fixed and the other floating. Ore feed is introduced into the gap between the rolls. The position of the floating roll can

FIGURE 4.11

High-pressure
grinding roll

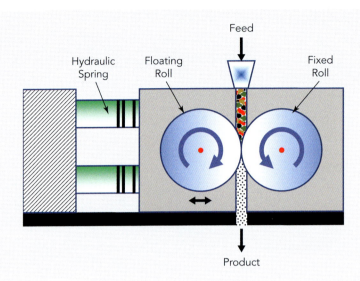

be adjusted so that the gap is slightly larger than the expected size of the largest diamond in the ore body. A hydraulic spring system maintains grinding pressure on the floating roll. The pressure and roll speed can be adjusted during the grinding to adapt to changing feed properties.

MORE ON HIGH-PRESSURE GRINDING MILLS

High-pressure grinding roll (HPGR) technology was originally developed for the cement industry. Diamond mines adopted the technology in the early 1980s for crushing kimberlite ore. HPGRs are now being used or considered for use in crushing gold and base metal ores where they would replace semiautogenous grinding (SAG) and autogenous grinding (AG) mills in a grinding circuit. Base and gold metal ores are typically harder than kimberlite.

Comminution in an HPGR is done almost completely by compression. This results in a product that has a higher percentage of fines than can be achieved with a SAG or AG mill where comminution is done by a combination of compression and shear. Coarse particles in the HPGR product exhibit extensive cracking, which reduces the amount of grinding work to be performed in a downstream ball mill.

HPGR units have a 6%–10% higher capital cost than SAG mills. One operating cost issue is wear of the roll surface (which is typically studded), particularly in gold and base metal ore processing. However, this is offset by the low cost of replacing wear surfaces, short equipment delivery times, and a high throughput rate. Energy costs of an HPGR are also significantly lower—most of the energy in a SAG or AG mill circuit is consumed moving the mill cylinder itself.

The product from the HPGR has a high percentage of fine particles. It is mixed with finely ground ferrosilicon (dense media) slurry at a density of approximately 2.65 g/cm³, close to the density of diamond. The resulting slurry is spun at high speeds in a cyclone, creating a density gradient in which lighter materials (kimberlite) rise to the top of the cyclone (the overflow) and are discarded as waste (or used as backfill in an underground mine). The higher-density minerals, including diamonds and garnets, form a diamond concentrate at the lower levels (the underflow) of the cyclone. Using a magnet, the ferrosilicon is recovered for reuse.

Next, the diamonds must be separated from the dense minerals. This is done using X-rays because diamonds fluoresce when subjected to X-rays. The X-ray separator system (Figure 4.12) acts on a thin stream of diamond concentrate particles accelerated off a moving belt into the air, where they encounter an intense beam of X-rays. A diamond fluoresces in the X-rays, activating a photomultiplier that triggers a jet of air, deflecting the diamond into a collector bin.

The waste stream is fed into one or two more X-ray separators to ensure complete diamond recovery.

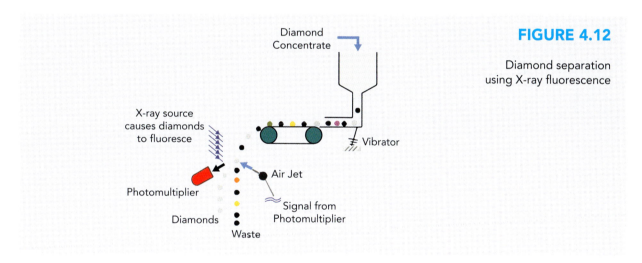

FIGURE 4.12

Diamond separation using X-ray fluorescence

AGGREGATES

Aggregates include sands and gravels, crushed stone (or crushed rock, or simply *crush*), and lightweight aggregates. Lightweight aggregates are shale and light volcanic rocks such as pumice or scoria (used in concrete). This section discusses sands, gravels, and crushed stone.

Natural resources of aggregates are extracted from quarries, gravel pits, and the sea (marine aggregates). A quarry is an open pit mine that produces rock materials for crushed aggregates. Loose material, such as sand and gravel, is extracted from a pit. Construction and demolition residues (e.g., concrete) are recycled to produce aggregates. Some aggregates are recycled to produce aggregates for other purposes. Manufactured aggregates are sourced from furnace slag (mostly silica) or from plastics.

Aggregates are used in the construction and maintenance of roads; the construction of buildings such as schools, hospitals, and administration facilities; and for other infrastructure such as harbors, waterways, and pipelines. The various products, materials, and end uses of aggregates are listed in Table 4.1.

TABLE 4.1

Aggregates: products, materials, and end uses

Products	Materials	End Uses
Crushed rock	*Concrete products:* Ready-mix, pre-cast, architectural	*Buildings:* Residential, commercial, public
Sand and gravel		
	Structural materials	Bridges, harbors, pipelines
Marine aggregates		
	Railway ballast	Roads, runways, railways, waterways
Recycled aggregates		
	Armor	
Manufactured aggregates		
	Asphalt	

Origin and Occurrence

Deposits of sand and gravel are produced by fluvial (river), glacial, or marine processes. The ubiquitous nature of these processes means that sand and gravel are widely distributed around the world.

Sources of crushed stone are any rock outcrop of sufficient size and quality to allow the economic production of crush. All rock types are used to produce crush, but the harder the rock is, the more energy is required to crush it, which leads to an increase in production costs.

Aggregate Processing

The processing of quarry aggregates is done by physical methods. Because the specifications for an aggregate can be demanding, the processing system has to be designed and operated to consistently achieve the specifications. For example, a concrete aggregate needs to be clean; consist of strong particles, perhaps of a certain shape; and free of chemicals, clay coatings, and other fines that could cause deterioration of the concrete.

The first step after extraction of the raw materials is to remove unwanted materials such as wood and clay. Clay particles are a particular concern. They are usually attached to or comingled with coarser sand and gravel particles but can be removed by washing and scrubbing. The feed material is combined with water, which removes some of the clay particles. Then it is fed to a system of intermeshing paddles that rotate to cause particle-on-particle attrition or scrubbing that removes the remaining clay coating. The clay is absorbed into the water and discharged.

The clean material then goes through various stages of crushing and screening to obtain the desired particle size or sizes. If necessary, some form of gravity separation or heavy media separation can be used to separate desirable lighter particles from heavier ones.

An example of sand and gravel pit operation is shown in Figure 4.13. The processing system on the left consists of a crusher, conveyors, and screens that are used to produce various types of aggregates stored on the nearby stockpiles.

Courtesy of Elphinstone Aggregates

FIGURE 4.13

A sand and gravel pit operation in British Columbia (Canada)

The Aggregates Business

The aggregates industry is enormous with operations in almost every country. The size of companies involved can vary from single-quarry operations to large international companies with multiple sources and large distribution networks. Global production of aggregates is not monitored uniformly, but available data show that more aggregates are extracted from the earth than any other metal or material (UNEP 2019; Bendixen et al. 2021).

In the developed world, aggregates are produced in quarries or production areas near cities and towns close to the end market. A relatively small amount of aggregates can occupy a large volume and weigh several tons so that transport by truck over distances greater than 80 km becomes cost prohibitive. Other transport modes, such as rail or water (barges), are commonly used, and this is one reason why larger aggregate companies also operate a distribution system.

A significant underlying driver of the demand for aggregates is increasing urbanization and population growth that results in increased demand for concrete used in buildings and infrastructure. Aggregate production is not monitored uniformly, but because almost 95% of aggregates are used in concrete (Zhong et al. 2022), a reasonable estimate of global aggregate production can be made using the more reliable data available for global cement production. Depending on the mix design, concrete consists of 6–10 t (metric tons) of sand and gravel for every metric ton of cement used (UNEP 2019). Thus, since global production of cement was approximately 4.2 Gt (gigatons) in 2022 (IEA 2023b), global sand and gravel production for use in concrete was about 33.6 Gt in 2022, assuming an average of 8 t of sand and gravel per metric ton of cement.

Cement production results in considerable carbon emissions, but steps have been taken by the construction industry to reduce demand for concrete. Consequently, demand for cement is expected to remain stable at 4.3 Gt until 2050, corresponding to 34.4 Gt of aggregates. However, until 2060, population growth and urbanization in Africa, India, and south Asia will put pressure on the supply of aggregates needed for construction (Zhong et al. 2022).

Crushed stone is the dominant source of manufactured aggregates, particularly in Europe where high-quality sand and gravel deposits are becoming depleted, and it is becoming more difficult to obtain permits for quarries and pits of any size. Similar circumstances in other parts of the world can be expected so that crushed stone will become the dominant source of manufactured aggregates in the future.

The aggregates industry is complex. Finding an economic source of aggregates is one challenge, and the efficient operation of a pit or quarry to ensure consistent high product quality is another. There is a certain glamour to the aggregate business (or any industrial mineral business) because a large part of it is dealing directly with customers—a producer of industrial minerals is closer to the customer than a metals producer.

For more details about industrial minerals, see *Industrial Minerals and Rocks* by Kogel and colleagues (2006).

Sand

Sources of sand for concrete production are large floodplains, glacial deposits, riverbeds, lakes, and shorelines. Desert sand cannot be used in concrete because the particles are too smooth to provide sufficient contact area for binding in a concrete mix. Seafloor sand contains chloride salts that can leach out of the concrete and cause corrosion of reinforcement or structural steel. This limits available sources.

However, the demand for sand is growing at an unsustainable rate (Bendixen et al. 2021; Zhong et al. 2022) so that available sand sources are rapidly depleted, especially in regions of high construction activity. Removal of excessive amounts of sand from these sources has a number of negative environmental and social consequences described by Bendixen et al. (2021).

Sand mining is not regulated in some jurisdictions and the lack of control and monitoring leads to unorganized operators who cause extensive damage, operate dangerously, and rapidly deplete resources. There have been many calls for regulated, formalized, and monitored sand mining operations and for schemes that promote sustainable use and reuse of sand (Bendixen et al. 2019; UNEP 2019).

REFERENCES

Bendixen, M., Best, J., Hackney, C., et al. 2019. Time is running out for sand. *Nature* 571:29–31.

Bendixen, M., Iverson, L.L., Best, J., et al. 2021. Sand, gravel, and UN Sustainable Development Goals: Conflicts, synergies, and pathways forward. *One Earth* 4:1095–1111.

Davis, W.J., and Kjarsgaard, B.A. 1997. A Rb-Sr isochron age for a kimberlite from the recently discovered Lac de Gras field, Slave Province, Northwest Canada. *Journal of Geology* 105:503–510.

EIA (U.S. Energy Information Administration). 2023. International coal and coke [reserves]. www.eia.gov/international/data/world/coal-and-coke/coal-reserves. Accessed August 2023.

Frolick, V. 1999. *Fire into Ice: Charles Fipke and the Great Diamond Hunt.* Vancouver, BC, Canada: Raincoast Books.

Hylland, R.L. 1995. Utah's state symbols. *Survey Notes* 27(2). https://geology.utah.gov/map-pub/survey-notes/glad-you-asked/state-symbols/. Accessed November 2023.

IEA (International Energy Agency). 2023a. Coal market update—July 2023. www.iea.org/reports/coal-market-update-july-2023. Accessed August 2023.

IEA (International Energy Agency). 2023b. Global cement production in the net zero scenario, 2010–2030. Cement: Activity. www.iea.org/energy-system/industry/cement. Accessed August 2023.

KGS (Kentucky Geological Survey). 2023. Kentucky's state mineral: Coal. www.uky.edu/KGS/education/state-coal.php. Accessed November 2023.

Kogel, J.E., Trivedi, N.C., Barker, J.M., et al., eds. 2006. *Industrial Minerals and Rocks*, 7th ed. Littleton, CO: SME.

Rygel, M.C. 2006. Sydney mines Point Aconi seam. https://commons.wikimedia.org/wiki/File:Sydney_Mines_Point_Aconi_Seam_038.JPG. Accessed August 2023.

Skinner, H.C.W. 2005. Biominerals. *Mineralogical Magazine* 69(5):621–641.

UNEP (U.N. Enviroment Programme). 2019. *Sand and Sustainability: Finding New Solutions for Environmental Governance of Global Sand Resources.* Geneva, Switzerland: UNEP. https://unepgrid.ch/en/resource/2AD54533. Accessed August 2023.

Zhong, X., Deetman, S., Tukker, A., et al. 2022. Increasing material efficiencies of buildings to address the global sand crisis. *Nature Sustainability* 5:389–392.

Mine Waste Management

Management of the waste streams produced by mining and processing minerals is a very important part of a mining operation. The reason these waste streams are significant is mainly because their mass is one to two orders of magnitude larger than the mass of the mineral product.

The material in this chapter describes the various mine waste streams and gives some idea of how they are managed and treated.

Mining and processing ore result in very large quantities of waste materials. The reason is that, as the typical ore grades mentioned in Chapter 1 suggest, a large part of what is extracted and processed is unwanted minerals. An example of the storage of the waste from mineral processing (called *tailings*) is shown in Figure 5.1.

Courtesy of Teck Resources Limited

FIGURE 5.1

Embankment at the downstream end of the tailings impoundment at Teck Highland Valley Copper Operations in British Columbia (Canada). The embankment is approximately 2.9 km long and about 165 m high. Water from the seepage pond (foreground) is pumped back into the impoundment.

Management of mine waste is a specialized topic. Because the waste is some combination of soil and rock, *geotechnical engineers* design the waste management systems and *environmental engineers* ensure that the systems have minimal impact on the air and water near the mine. There are so many aspects and details associated with mine waste management that it is impossible for one person to be an expert in all aspects of mine waste management.

There are four waste streams from a mine:

1. Tailings—the waste from the mineral processing operation
2. Waste rock—mined rock of low to zero grade, present in large amounts at an open pit mine and in very small amounts at an underground mine
3. Wastewater—excess water from a variety of sources such as the pit, waste dump, and tailings storage facility
4. Acid rock drainage—low-pH, metal laden, sulfate-rich drainage that is the result of oxidation of sulfide minerals

Each of these waste streams must be managed during the life of the mine and after the mine closes. A management plan for this entire period must be in place before the mine obtains a permit to operate. Designing the management system is a significant engineering challenge.

TAILINGS

Tailings are finely ground rock from the concentrator/processing plant. The particles are less than 0.1 mm in diameter, the size of beach sand or silt. Figure 5.2 shows wet tailings, a mixture of particles and water, being discharged into a tailings impoundment.

FIGURE 5.2

Wet tailings slurry being discharged into a tailings impoundment

Courtesy of Jon Engels, www.tailings.info

When tailings exit the concentrator, they are in the form of a slurry composed of sand and silt particles and water. They can be disposed of as wet tailings (*slurry*), *thickened* tailings, a *paste*, or *filtered* (dewatered) tailings. These disposal methods are described in the following subsections. Table 5.1 lists the types of tailings and compares their consistency with well-known materials. The behavior of the different forms of tailings is almost completely governed by their solids content, the ratio between the mass of solids to the total mass of solids and water. As the solids content of tailings increases, more energy is required to pump it. Ultimately, if the solids content exceeds 85%, the tailings cannot be pumped and must be transported by conveyor or truck.

Wet Tailings Disposal

The water content of the slurry that exits from the concentrator is high, and this increases its volume. To reduce the volume, some water is removed from the tailings using a thickener (see the "Physical Separation Methods" section in Chapter 3). The water is recycled to the processing plant.

Tailings Form	Consistency	Solids Content, %	Pump or Transport
Slurry	Sand and water	30–55	
Thickened	Milkshake	50–70	Can be pumped
Paste	Toothpaste	70–85	
Wet filtered	Wet sand	>85	Must be transported by conveyor or truck
Dry filtered	Moist sand	>85	

TABLE 5.1

Comparison of tailings

Wet tailings are deposited behind an embankment. It is important to distinguish a *tailings embankment* from a *tailings impoundment*. A tailings impoundment is shown in Figure 5.3. The embankment is the structure that retains the tailings and the ponded water, whereas the impoundment includes the embankment, disposed tailings, and ponded water. (A *tailings storage facility* includes the impoundment as well as infrastructure and equipment to handle tailings such as roads, pipes, pumps, and cyclones.) How these components of the impoundment are constructed or formed is described in this subsection.

The embankment is built of some combination of waste rock or the coarse fraction of the tailings that is obtained using cyclones. A *filter* consisting of layers of soil or tailings is often placed on the upstream side. There is a gradation in particle sizes of the filter layers that allows fluid to flow through the filter but prevents excessive migration of tailings particles. Filters might also be placed within an embankment

FIGURE 5.3

Components of a tailings impoundment

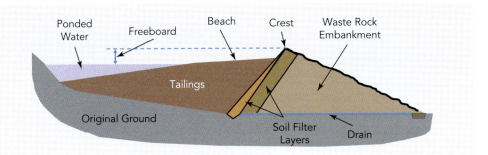

FIGURE 5.4

Wet tailings discharge into impoundment at Bagdad mine, Arizona

to prevent movement of fine particles through adjacent zones of coarse particles that, if excessive, is called *internal erosion*.

Wet tailings impoundments can be almost any shape as long as the embankment is able to contain the tailings and water. There are many possible configurations, depending mostly on topography and availability of materials for construction (see Vick 1990). As an example, Figure 5.1 shows the downstream part of the impoundment at the Highland Valley mine in south central British Columbia, while Figure 5.4 shows the impoundment at the Bagdad copper mine in Arizona (United States), a much drier and hotter climate where the evaporation rate is significant.

The tailings are deposited from spigots, as shown in Figure 5.2. Coarse particles segregate and settle close to the spigots, forming a *beach*, while the fines are transported and settle furthest away from the spigot. One to several spigots may be used depending on the volume of tailings being handled. The fewer the number of spigots, the more the spigots will have to be moved to avoid excessive accumulation at one location.

As the tailings in the impoundment settle, consolidation of the particles occurs, reducing the pore space between the particles and forcing water out to form a *pond*. If necessary, water may be reclaimed by pumping it from the tailings pond.

The distance between the pond elevation and the crest elevation is called the *freeboard* (Figure 5.3). A nonzero freeboard is maintained to minimize pore water pressures in the tailings, which would place an unwanted load on the embankment. Small or zero freeboard might occur because of excessive rainfall or inflow, and the embankment is designed to withstand this, but it cannot be a permanent condition. The pond water would eventually have to be treated, if necessary, and released.

Tailings Embankment Designs

There are basically three ways to construct an embankment for wet tailings disposal. One is the *upstream method*, illustrated in Figure 5.5. A starter dike is constructed from the cycloned coarse fraction of the tailings or waste rock, and tailings are deposited behind it. As mining and processing proceeds, the impoundment volume must increase. This is done by constructing additional *raises* using coarse tailings and founded on the beach formed behind the previous raise. This moves the centerline of the embankment upstream, hence the name of the design. Depending on the volume of the impoundment and height of the raise, each raise can provide enough capacity for 1–3 years of tailings storage.

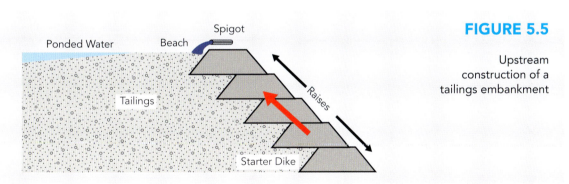

FIGURE 5.5

Upstream construction of a tailings embankment

A large freeboard is maintained behind an upstream-constructed embankment to allow the beach to dry and minimize the pressure that water in the tailings would impose against the embankment. The dried tailings are also compacted and often used to construct the next raise. A cover of waste rock is placed on the downstream face.

Upstream design uses the least amount of material for construction, thus reducing costs, but each of the raises is partly founded on weak and saturated beach tailings—an unstable foundation—especially in the event of an earthquake. Although compaction of the beach improves the foundation stability, upstream construction is not allowed in seismically active areas. Franks et al. (2021) found that tailings embankments built with an upstream design exhibited the largest percentage of stability issues.

Almost the exact opposite of the upstream method is the *downstream method*, illustrated in Figure 5.6. Downstream construction avoids the instability problems associated with upstream construction because each raise is founded on stable material in front of the previous raise, thus moving the centerline of the embankment downstream. Stability of the downstream design can be further enhanced by the use of internal drainage made of porous sand, which ensures that the water level in the embankment remains low. The filter layers of silt or clay in the upstream side of the embankment also keep water out of the embankment.

Embankments built with a downstream design have exhibited a relatively lower percentage of stability issues (Franks et al. 2021). However, downstream construction requires the largest amount of material of any of the three embankment designs.

In between upstream and downstream construction in terms of the amount of material required is *centerline construction*, shown in Figure 5.7. Filter layers and internal drains can also be used in embankments with centerline design.

FIGURE 5.6

Downstream construction of a tailings embankment

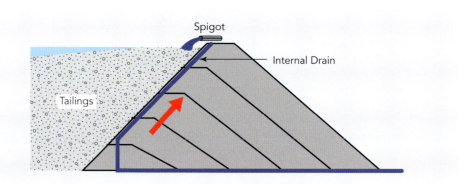

FIGURE 5.7

Centerline construction of a tailings embankment

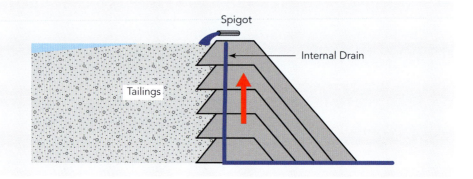

The cross section of any tailings embankment is often much more complex than the simplified diagrams shown here. The exact form depends on the availability of material, details of the construction sequence, and the geometry of the impoundment.

The distance from the processing plant to the tailings impoundment and the elevation difference between these two locations has a significant effect on capital and operating costs of tailings storage. The pipelines used to transfer tailings to the impoundment and to reclaim water can cost from US$300,000 to US$400,000 per kilometer, and, of course, the longer the distance, the greater the pumping costs. However, constraints on location of the impoundment arise if the volume of tailings is large or if topography does not allow location of an impoundment close to the plant.

Tailings Water Balance

Water flows into and out of a wet tailings impoundment. The various flows are shown in Figure 5.8. Whether the amount of water that enters the impoundment is equal to the amount that comes out is closely monitored using data from measurement devices located within and around the impoundment. If there is a significant discrepancy (greater than the combined measurement errors), then there may be a leak, which must be repaired before processing can continue.

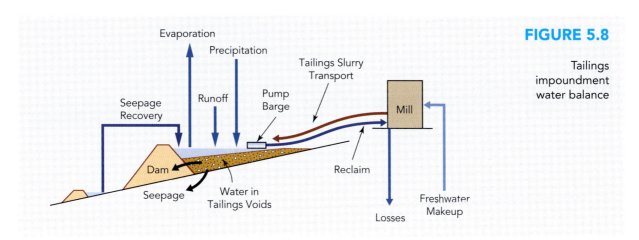

FIGURE 5.8

Tailings impoundment water balance

Thickened, Paste, and Filtered Tailings

Partial or complete removal of water from tailings makes the tailings much easier to handle and allows more possibilities for storage. Thickened tailings result when the tailings from the plant are pumped into a cone-shaped thickener. The result is a slurry with a water content between 30% and 50% that is fluid enough to be pumped (Figure 5.9). The resulting tailings footprint is much smaller and more stable than wet tailings.

HISTORY OF TAILINGS EMBANKMENT FAILURES

The following graph shows the number of recorded failures of tailings embankments worldwide for each decade from 1910 to 2019. There is a total of 124 failures where the embankment failed leading to a discharge of tailings and another 200 failures where the embankment did not fail but discharge occurred. The failures are classified in terms of the volume D of discharged tailings as follows:

Classification	Discharge Volume, D	No. of Failures
Very serious	$D \geq 1 \text{ Mm}^3$	53
Serious	$0.1 \text{ Mm}^3 \leq D < 1 \text{ Mm}^3$	71
Other	$D < 0.1 \text{ Mm}^3$	200

Source: Center for Science in Public Participation 2023

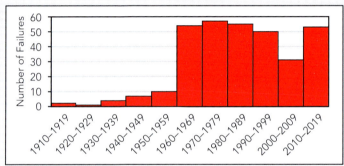

Source: Center for Science in Public Participation 2023

The low failure rate during the early 20th century is likely because of the small size of tailings embankments and a lack of reporting. There is a marked increase in failure rate after 1960, which is believed to be caused by lower ore grades leading to increased production of tailings, much larger impoundments, and higher embankments. Enabled by the larger sizes of earthmoving equipment, tailings were placed without a full understanding of the behavior of such large masses. Faulty design, poor construction control, and lack of monitoring during operation are other contributors. The decrease in the number of failures from 1990 to 2009 may be due to lack of reporting or failure classification errors.

Concerns about tailings impoundment safety increased after 2010, mainly because of the very serious failures at Mount Polley in 2014, Samarco in 2015, and Brumadinho in 2019. It became evident that there were no standards for tailings management, that recommendations made after past failures were not being followed, and that impacts of tailings storage to communities were significant. Consequently, 100 investors with very large holdings in mining companies formed a coalition (COE 2019). Its first task was to ask the industry to disclose all engineering and governance aspects of each of their tailings facilities. After about 1½ years of consultations with experts from industry, consulting companies, academia, and community stakeholders, the first global standard on tailings management was developed and launched in 2020 (GTR 2020).

FIGURE 5.9

Thickened tailings

Courtesy of Frank Palkovits, Kovit Engineering

FIGURE 5.10

Discharge of paste iron ore tailings from thickener

Courtesy of WesTech Engineering Inc.

Using a deep cone thickener, tailings can be thickened to a high solids content and the result is called *paste*, shown in Figure 5.10. Some underground mines add cement to paste to provide strength, and the resulting material is used as backfill.

Tailings may be completely dewatered using a filter press, shown in Figure 5.11a. The tailings are placed between a series of hard plastic plates, and pressure is applied to force the water out of the tailings. A filter cloth traps the particles between the plates. When the water is forced out, the plates open and the resulting *filter cake* drops onto a conveyor belt. Figure 5.11b shows a filter press at full scale.

Dewatered tailings can be deposited in any convenient stable formation, known as a *dry stack*. An example is shown in Figure 5.12.

If thickened, paste, or filtered tailings are so much easier to handle and reduce the footprint of an operation, why not use them in all cases? Mostly, because it is expensive. Filtered tailings are economically justified in a very dry climate where all or most of the water must be reclaimed due to its scarcity. In a case where

FIGURE 5.11

Filter press
operation and
filter press

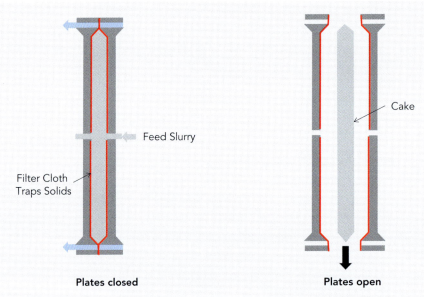

Feed Slurry

Filter Cloth
Traps Solids

Cake

Plates closed

Plates open

(a) Filter press operation

(b) Filter press showing the polypropylene plates and filter cloths between them
(Courtesy of McLanahan Corporation)

FIGURE 5.12

Dry tailings
disposal using
haul trucks

Courtesy of Jon Engels, www.tailings.info

topography or location make it desirable to avoid wet tailings disposal, the costs of thickened or dry tailings can be justified if the ore is high grade or inexpensive to mine. However, it is also important to note that filtered tailings are not completely devoid of water. In a climate where rainfall occurs, filtered tailings would absorb any water like a sponge, and some facility (e.g., a berm) must be constructed to control seepage from the filtered tailings. In a dry, hot climate, such as that shown in Figure 5.12, the water in the filtered tailings would evaporate.

Another issue with thickened, paste, and filtered tailings is variability of the solids content and of the flow rate of the tailings stream because of changes in the processing plant or from the occurrence of a storm or excessive runoff. For example, if a tailings disposal facility is designed for a particular high solids content and the processing plant produces tailings with lower solids content (e.g., more like a slurry), then it will be difficult to manage the lower-density tailings. Flexibility to handle such variability can be designed into the disposal system, but it comes at a cost.

WASTE ROCK

Waste rock is nonmineralized rock and/or low-grade ore. A lot of waste rock is produced at an open pit mine and is stored in large *waste dumps* (or *waste piles*). Very little waste rock is produced at an underground mine, and most of it is often stored underground.

Waste dumps are enormous masses produced by dumping waste rock in a controlled manner over a large area. Figure 5.13a shows a haul truck dumping waste rock and illustrates the range of particle sizes present, typically from clay size to 2 m. An example of the large waste dump that results is shown in Figure 5.13b. Waste dump operation and management are affected by the mine plan, that is, the geometrical relationship between the waste and ore in the mine and the strip ratio.

A waste rock dump may seem to be a barren mass of rock. However, the interior of a dump is very active, physically and chemically (Figure 5.14). Segregation after dumping causes coarse particles to settle near the bottom of the dump face and finer particles to accumulate near the top. Naturally occurring bacteria catalyze the oxidation of sulfide minerals in the dump, leading to the generation of acid and metals in solution. Water from precipitation infiltrates the dump, and its flow must be controlled to maintain stability and prevent acid or metal contamination of surface water or groundwater.

Probably the most interesting phenomenon in a waste dump is the result of heat generated by oxidation of the sulfides. Cold air enters the dump through the coarse

FIGURE 5.13

Waste rock

(a) Truck dumping waste
(Courtesy of L. Smith, The Groundwater Project)

(b) Large waste dump
(Courtesy of J. Caldwell)

FIGURE 5.14

Cross section of a
waste rock dump
showing chemical and
physical processes

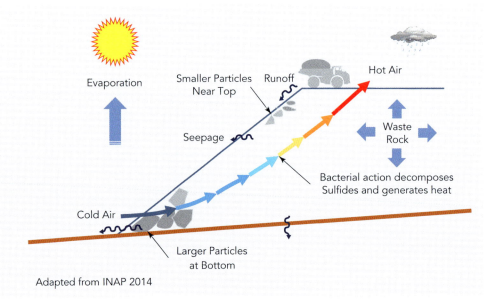

Evaporation

Smaller Particles
Near Top

Runoff

Hot Air

Seepage

Waste
Rock

Cold Air

Bacterial action decomposes
Sulfides and generates heat

Larger Particles
at Bottom

Adapted from INAP 2014

material at the bottom; becomes heated, resulting in an increase in its moisture content; and vents as an updraft of hot, damp air from the top surface of the dump. The venting may be visible if the ambient temperature is cold enough.

HOW MANY TONS OF TAILINGS AND WASTE?

Consider a copper mine with 0.5% grade mining 170,000 t (metric tons) per day, 360 days per year, for 20 years. That is 1,224 Mt (million metric tons) of waste and ore. Suppose the strip ratio of waste/ore = 0.5. Thus, waste + ore = 1,224 Mt and waste = ore × 0.5. This gives 816 Mt of ore and 408 Mt of waste.

Suppose 27% copper concentrate is produced, and the recovery in the concentrator is 90%. A simple mass balance of copper gives

$$K \text{ metric tons ore} \times 0.005 \times 0.9 = 1 \text{ t concentrate} \times 0.27$$

from which K = 60 t of ore to produce 1 t of concentrate. Thus, from 816 Mt of ore 816/60 = 13.6 Mt of concentrate will be produced. This leaves 816 − 13.6 = 802.4 Mt of tailings.

In fact, 8 Mt of concrete was used to build the Hoover Dam on the Colorado River at the border between Arizona and Nevada (United States).

WASTEWATER

Wastewater consists of liquid wastes from processing such as flotation reagents, solvent extraction and electrowinning solutions, and acids or cyanide used in leaching. Wastewater from a mine may contain ammonia from explosives or metal-contaminated groundwater.

Wastewater must always be treated before release to the environment. The best strategy is to minimize the production of wastewater, but this usually leads to higher costs. It could be contained, as in behind a dam, but this is difficult and risky. Sometimes the water can be treated and reused in the processing plant.

ACID ROCK DRAINAGE

When sulfide minerals are exposed to air and rain, a chemical reaction occurs that breaks down (oxidizes) the sulfide and makes the water acidic. This is generally known as *acid drainage* but is also called *acid rock drainage* (ARD) or *acid mine drainage* (AMD). Metal elements bound in the sulfide are also released into solution, resulting in metal contamination.

Owing to its relative abundance, pyrite is the largest contributor to the generation of ARD. A summary chemical reaction for the oxidation of pyrite and generation of acid is

pyrite	+	oxygen	+	water	$\rightarrow$	sulfuric acid	+	iron hydroxide "yellow boy"
$4FeS_2$	+	$15O_2$	+	$14H_2O$	$\rightarrow$	$8H_2SO_4$	+	$4Fe(OH)_3$

Similar reactions occur for other metal sulfides. The result is the generation of acid and metal contamination of the water.

The most common way to treat acid drainage is to neutralize it with lime (calcium hydroxide). However, this is expensive, and considerably more lime than that required to neutralize the water must be added to completely precipitate the dissolved metals and obtain water with a metal content low enough to meet common standards. In addition, the calcium sulfate sludge produced by lime neutralization is difficult to manage because it is saturated and must be constantly agitated or recirculated to prevent settlement and pipe clogging.

Prevention of acid generation is a better option. This requires keeping rock that can produce acid—that is, rock with high concentrations of sulfides—underwater, which prevents exposure to oxygen so that the preceding reaction will not occur.

Although acid generation is a natural process, it is moderated by the presence of carbonate minerals that neutralize the acid. Mining, construction, and waste disposal activities expose large surface areas of rock containing sulfides to water and oxygen, and the amount of carbonate minerals present is not sufficient to neutralize the acid generated.

RIVERS OF ACID

Some rivers in the Iberian Peninsula and Latin America are naturally acidic because of their proximity to sulfide mineralization. The Rio Tinto in southern Spain is one example, where the river water is acidic and its red color (*tinto* means "red" in Spanish) is caused by metal precipitation. Mining activities near the Rio Tinto have occurred over several thousand years in the surrounding area and likely increased the acidity, but there are massive deposits of metal hydroxides and sulfates in the area that predate any mining activity, suggesting that mining did not initiate the acidity.

Three rivers are named Rio Agrio (*agrio* means "sour" in Spanish): one in Spain, another in Argentina, and another in Costa Rica. The water in each exhibits color, turbidity, and acidity.

An example of ARD and its treatment at the Equity Silver mine site (British Columbia, Canada) is shown in Figure 5.15. The iron hydroxide, or "yellow boy," is visible in the collection pond in Figure 5.15a. The mine was in operation from 1980 to 1994 and is currently under care and maintenance.

(a) Collection pond with "yellow boy"

(b) Treatment plant

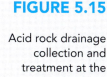

FIGURE 5.15

Acid rock drainage collection and treatment at the Equity Silver mine site (British Columbia)

(c) Discharge of treated water

All images courtesy of Newmont Mining Corporation

The original owner of the mine was Placer Dome, but in 2006, Placer Dome was bought by Barrick Gold, who spun off some mining assets that were acquired by Goldcorp. As part of the transaction, Goldcorp assumed responsibility for the monitoring and treatment at the site. In 2019, Goldcorp was acquired by Newmont Mining Corporation who assumed responsibility.

In 2012, just over 4,000 t of lime were used to treat the ARD. The average annual operating cost for the property from 2009 to 2013 was $2,065,000, of which the major costs were lime, power, and wages. This is a good example of the potential lingering liabilities of mine waste; that is, they do not go away. ARD from waste dumps at the mine will require monitoring and treatment for the foreseeable future.

When the mine was built, little was known about ARD. Nowadays, mine waste storage facilities would be designed to avoid or mitigate the generation of ARD (Verburg 2011).

MINE WASTE TREATMENT AND DISPOSAL METHODS

Over the past few decades, engineers and scientists have developed and improved methods for treating and managing mine waste using physical and chemical techniques. A few of these methods are described in this section.

Reclamation

Reclamation is a set of operations designed to make a mine waste disposal site blend in with the surroundings and possibly become usable land. An example of reclamation of a gold mine tailings impoundment at the Waihi gold mine in New Zealand

FIGURE 5.16

Waihi gold mine, New Zealand

(Courtesy of K. Wilson/ Newmont Waihi Gold)

(a) January 2003—Tailings impoundment during operation

(b) February 2011—Reclaimed impoundment following mine closure

is shown in Figure 5.16. The reclamation process occurred during and after the mine was in operation.

In general, reclamation must be carefully planned and begins soon after the mine begins operation. As an example, planting simple grasses on tailings is done in stages. First, plants that grow in barren soil are planted. Next, these plants are allowed or caused to decompose to form an organic layer. This might be done several times to form a thick layer. Finally, the desired grasses are planted.

Placement of Tailings and Waste into Empty Pit

Returning the tailings and waste to an empty pit is one way to avoid some of the long-term liabilities associated with disposal of tailings and waste on land, especially those related to water draining from tailings and waste. Moving such large amounts of material after they have been placed is expensive, but this expense must be compared to the much larger expense of managing the waste and treating the water draining from it for an indefinite period.

At the Kidston Gold Mines in Australia, it was decided to thicken the tailings and deposit them into one of two pits. (The two pits are now part of a pumped hydroelectric scheme [Genex, n.d.].) Waste rock was also dumped into the pit and allowed to mix with the tailings (Figure 5.17).

FIGURE 5.17

Disposal of tailings and waste rock into pit at left at Kidston Gold Mines, Australia

Source: Williams 2021

The flow of water from the mix of tailings and waste in the pit is the main concern in this approach to waste disposal. The water may be partially removed, as in thickening, or an impermeable barrier may be placed at the bottom and sides of the pit.

Sequestration of Acid-Generating Waste

If a sufficiently accurate ore-body model is available, the distribution of waste rock containing high concentrations of sulfide will be known. Rock with such high-sulfide concentrations can be expected to generate acid. The high-sulfide waste can be separated from the low-sulfide waste as part of the mine plan, and then the dumping of waste can be planned such that the high-sulfide waste is kept saturated and covered by the low-sulfide waste. Many variations of this theme are possible. One is illustrated in Figure 5.18. Note that such a disposal scheme could not be contemplated if the tailings were filtered.

FIGURE 5.18

Sequestration and covering of high-sulfide waste

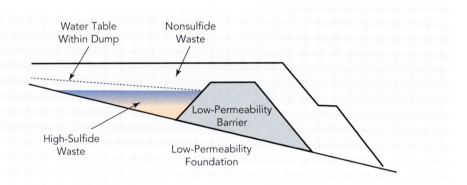

Co-Disposal

Mine waste can be divided into fine-grained material, such as tailings, and coarse-grained material, such as waste. The conventional approach is to dispose of these two waste streams separately. The voids between particles in fine-grained tailings are filled with water, which contributes to their instability (an angle of repose equal to zero). Waste is stronger (has a higher angle of repose), but the voids between the particles in waste are filled with air, which allows oxygen and water to flow relatively easily, leading to acid generation and metal contamination of the water. Co-disposal is where the tailings are forced to settle in the voids in the coarse waste, resulting in a new material that is strong and quite impermeable. Results of co-disposal tests at the Porgera gold mine in Papua New Guinea are shown in Figure 5.19.

Co-disposal has been applied at some mines in Australia (Gowan et al. 2010). The tailings and waste can be mixed either at the disposal site or before transport to the disposal site. Gowan et al. describe several methods for mixing by pumping and placing the materials. Two other advantages of co-disposal are high water return to the processing plant and lower costs of pumping compared to trucking waste to a dump.

It is actually difficult (and expensive) to mix these two materials—the expectation is that the tailings will flow into the voids of the waste once they are pumped over

FIGURE 5.19

Co-disposal of tailings
and waste rock
(Courtesy of G.W. Wilson,
University of Alberta)

(a) Mixing of tailings and waste rock   (b) Consolidated mixture

the waste. Whether this happens depends mostly on the difference between the size distributions of particles within the tailings and the waste, and this can vary depending on how the concentrator operates. Fines in the waste will block the flow of tailings to the voids in the waste. In the extreme, if simple placement of the tailings does not work, equipment similar to cement mixers may be needed.

A flexible co-disposal method must be designed before mining begins to ensure that the mixing will occur at all stages. Logistics and the possible variability of particle sizes would make it difficult and expensive to mix tailings in an existing impoundment with rock in an existing waste rock dump for co-disposal.

Removal of Metals from Mine Wastewater

Sulfate-Reducing Bacteria

Mine wastewater often contains metals and may or may not be acidic. It can be treated using microorganisms called *sulfate-reducing bacteria* (SRB). The SRB catalyze the reduction of metal sulfates present in mine wastewater, which results in the precipitation of metal sulfides. SRB exist naturally but need a carbon source to become active. Figure 5.20 illustrates the process.

The sulfides can be isolated by sequestration in a landfill. However, sometimes the sulfides contain sufficient amounts of metal that it is profitable to process them to obtain pure metals.

Constructed Wetlands

Acidic mine wastewater containing metals can also be treated in a passive manner using constructed wetlands, which are built to encourage biological and chemical reactions that neutralize acid and precipitate metals. A cross section of a typical

FIGURE 5.20

System diagram of a biological process used to treat mine wastewater

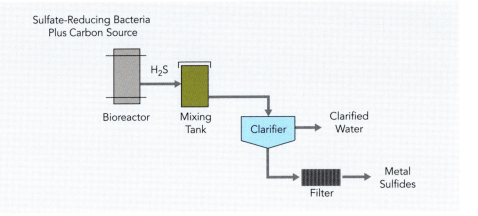

Before introduction to the wetland, contaminated water is placed in a pond where suspended particles settle. Several processes occur simultaneously in a wetland (Ford 2003). Aquatic plants such as cattails transport oxygen from their roots to the water, which aids oxidation of ferrous iron (Fe^{2+}) or other metal ions in solution, leading to precipitation of a metal hydroxide [e.g., $Fe(OH)_3$] in the organic layer. This is an *aerobic* process; that is, it involves oxygen. *Anaerobic* biological processes involving SRB also occur in the organic layer, leading to reduction of sulfates and precipitation of metal sulfides. Precipitation of the sulfide minerals prevents them from covering the surfaces of the underlying limestone particles and interfering with neutralization of the acid.

The organic layer can usually absorb metals for about 20 years, after which it can be removed for disposal in a landfill. It is not considered hazardous waste. More often, it is simply left in place and monitored.

Passive treatment systems have been applied to the treatment of municipal waste and landfill effluent for some time. Their application to the treatment of AMD is

wetland is shown in Figure 5.21. It consists of a thick (1–2 m) layer of compost under 10–30 cm of slow-flowing water containing aquatic plants. An underlying layer of limestone may be present to neutralize water of high acidity.

FIGURE 5.21

Cross section of a typical wetland

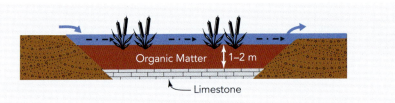

FROM SULFATES TO SULFIDES USING SRB

The energy required by sulfate-reducing bacteria (SRB) metabolism is derived by sulfate reduction, not by oxidation of sulfides as with other bacteria. SRB require an organic carbon source (as food) and some growth substrate for attachment given that the bacteria cannot survive in open water or any aerobic (high-oxygen) environment.

The biological names of common SRBs are *Desulfovibrio* and *Desulfotomaculum*. They can tolerate wide variations in acidity and temperature.

If sucrose (sugar) is the carbon source, the chemical reactions are as follows:

sucrose + sulfate + acid + water → hydrogen sulfide + carbonic acid

$$C_{12}H_{22}O_{11} + 6SO_4^{2-} + 12H^+ + H_2O \rightarrow 6H_2S(aq) + 12H_2CO_3(aq)$$

metal ion + hydrogen sulfide → metal sulfide + hydrogen ion

$$Me^{2+} + H_2S(aq) \rightarrow MeS(solid) + 2H^+$$

SRB cause sulfide minerals to precipitate in marine sediments, wetlands, lake sediments, or wherever there are sources of metal ions, sulfate ions, and carbon.

relatively recent and is an area of active research. A significant problem is understanding the role of bacteria and how bacterial colonies evolve within the system. With some design adaptations to particular local conditions, passive systems can work effectively in cold climates (Ness et al. 2014).

REFERENCES

Center for Science in Public Participation. 2023. TSF failures from 1915. www.csp2.org /tsf-failures-from-1915. Accessed July 2023.

COE (Church of England). 2019. The Investor Mining and Tailings Safety Initiative. www .churchofengland.org/about/leadership-and-governance/church-england-pensions-board /pensions-board-investments/investor-1. Accessed November 2023.

Ford, K.L. 2003. *Passive Treatment Systems for Acid Mine Drainage.* Technical Note 409. BLM/ST/ST-02/001+3596. Denver, CO: Bureau of Land Management, National Technology Center.

Franks, D.M., Stringer, M., Torres-Cruz, L.A., et al. 2021. Tailings facility disclosures reveal stability risks. *Scientific Reports* 11:5353.

Genex. n.d. 250 MW Kidston pumped storage hydro project. https://genexpower.com .au/250mw-kidston-pumped-storage-hydro-project/. Accessed July 2023.

Gowan, M., Lee, M., and Williams, D.J. 2010. Co-disposal techniques that may mitigate risks associated with storage and management of potentially acid generating wastes. In *Mine Waste 2010: Proceedings of the First International Seminar on the Reduction of Risk in the Management of Tailings and Mine Waste*. Edited by R. Jewell and A.B. Fourie. Perth: Australian Centre for Geomechanics. pp. 389–404. doi:10.36487 /ACG_rep/1008_33_Gowan.

GTR (Global Tailings Review). 2020. *Global Industry Standard on Tailings Management*. https://globaltailingsreview.org/wp-content/uploads/2020/08/global-industry -standard_EN.pdf. Accessed November 2023.

INAP (International Network for Acid Prevention). 2014. *Global Acid Rock Drainage (GARD) Guide*. Melbourne, Victoria, Australia: INAP. www.gardguide.com. Accessed November 2023.

Ness, I., Janin, A., and Stewart, K. 2014. *Passive Treatment of Mine Impacted Water in Cold Climates: A Review*. Whitehorse, Canada: Yukon Research Centre, Yukon College. https://casinomining.com/_resources/Passive_treatments_review_-_Cold_Climate_ -_YRC2014.pdf. Accessed November 2023.

Verburg, R. 2011. Mitigating acid rock drainage. In *SME Mining Engineering Handbook*, 3rd ed. Edited by P. Darling. Englewood, CO: SME. pp. 1721–1732.

Vick, S.G. 1990. *Planning, Design and Analysis of Tailings Dams*. Richmond, BC, Canada: BiTech.

Williams, D.J. 2021. Lessons from tailings dam failures—Where to go from here? *Minerals* 11(8):853. doi:10.3390/min11080853.

Mining, Society, and the Environment

Mining is done by people, near communities of people, within a social system, and in a physical environment. To fully understand how mining works, it is essential to understand how mining affects people, communities, society in general, and the environment.

This chapter introduces the essential aspects of human resources, health and safety, mining and society, and the environmental effects of mining. Any one of these topics could be (and is in some cases) the subject of an entire book. Their importance has grown over the last decade or more. They illustrate the breadth of the mineral resources industry, the variety of people associated with it, and the different professions within it.

Mining is a multidimensional subject. The references provide further details.

HUMAN RESOURCES

The mining industry requires people with all kinds of skills and education. However, shortages of qualified persons who can lead, manage, or operate mining businesses are a significant concern. In particular, there are shortages of experienced specialists such as mine planners, processing engineers, and mining engineers with digital skills (Abenov et al. 2023). (The mining industry is undergoing a digital transformation in which, for example, sensors on operating equipment provide large amounts of data that can be analyzed to detect patterns and allow informed decisions concerning operation or maintenance. See Chapter 8.) However, at the post-secondary level, it is difficult to attract young people, regardless of gender or ethnicity, to programs in mineral resources and mining (MIHRC 2023).

These issues represent significant challenges that affect the future productivity of the industry. They are the result of a complex set of factors that are not country specific. The career expectations of Gen Z (born 1995–2005) are typically based on environmental or social values and work–life balance. Newer generations will likely have similar desires and expectations. Unfortunately, mining has a reputation for causing environmental damage and disrupting the economic and social fabric of communities and even entire countries. Work schedules in remote mine sites (e.g., 2 weeks on, 2 weeks off) restrict work–life balance.

Surveys of students in North American and Australian universities reveal a significant lack of knowledge or understanding of what the mining industry is (Banta et al. 2021; Barton et al. 2021). The sources of metals and materials are also unknown. This is a fundamental problem, but some depictions of a career in mining are not helpful and give the wrong impression. Pictures of people in hard hats, goggles, ear protectors, overalls, steel-toed boots, and two-way radios standing next to a large piece of machinery situated on a pile of broken rock suggest harsh, possibly dangerous, working conditions or that a career in mining will involve actually operating the machinery. More importantly, such depictions do not convey the breadth of career possibilities in the industry.

These negative perceptions are possibly the main reasons why a career in mining does not appeal to young people. The result is a decrease in the number of graduates from mining engineering programs. Since 2016, the number of graduates of mining engineering programs has decreased by 40% in the United States and 60% in Australia (Abenov et al. 2023). In Canada, the decrease since 2016 is 45%.

To combat the low supply of engineers, particularly those having data science skills, some companies hire engineers from other engineering or science disciplines and provide them with formal or on-the-job training in mining or mineral processing operations. Some mining companies have entered into agreements with universities or technical schools to support education and training programs for mining professionals and workers. These programs are most successful in supplying the necessary technical trades for the mining workforce.

More support is needed for in-house or external educational programs that could deal with the potential skill and leadership shortages in the industry. Collaboration between educational institutions and the industry are essential to ensure that programs are relevant, interesting, and attract the best and brightest. The industry is changing (see Chapter 8), and any educational program should take this into account.

Women in Mining

Women are underrepresented in the mining workforce. As of 2016, between 8% and 17% of the global mining workforce were women (Fernandez-Stark et al. 2019), but the majority of these positions are entry level in equipment operation or administrative roles. The proportion of women decreases rapidly from entry-level to management and senior executive roles (Ellix et al. 2021). A career in mining offers women a variety of work, opportunities for advancement, and large salaries. However, surveys of women in the industry conducted in Australia, Brazil, Canada, and the United States reveal that many women want to leave because the work was found to be not interesting or challenging, there were no opportunities for growth, and support or mentoring were lacking (Ellix et al. 2021).

These negative circumstances can be difficult to mitigate because of the way the industry has evolved. Technology changes may provide more opportunities for women, but ongoing support from senior management to include women is essential. For example, a mine in Australia made a determined effort to support development of a gender-balanced workplace. Its workforce is now 40% women. Managers had to redesign entry-level positions so that women with a variety of backgrounds and education could be hired and trained in the newer technologies used at the mine (Denison and Pringle 2023).

SAFETY

There are many definitions of *safety*, but basically it is a work environment that avoids or mitigates conditions that can cause harm to working personnel. The question then becomes "How does one provide such an environment?" This is where the link to the technical aspects of an operation enters. However, what makes the issue more interesting and challenging is that the environment includes humans whose behavior plays a significant role in safety, and therefore, the workplace must be organized and managed in a particular way.

Hazard and Risk

Hazard and *risk* are important concepts in all aspects of mining operations, including workplace safety management. Often, the terms are used interchangeably, but this can lead to confusion. A general definition of *hazard* is

> A source of potential harm that could be caused by a dangerous or damaging event or sequence of events

The event may or may not have consequences nor any safety implications. Risk is the combination of a hazardous event and its consequences. Quantitatively, if P is the probability of a hazardous event and C is a measure of its consequences, then

$$\text{risk} = P \times C$$

Thus, if the probability of a hazardous event is 0.1 and the consequence is a US$1 million loss, then a measure of the risk is US$100,000. However, in many cases, such a measure of risk cannot be made because the probability cannot be estimated and the consequences are unquantifiable.

Risk Management

Safety can also be defined as the control of hazards to achieve an acceptable level of risk. This means that hazards must be identified and an acceptable level of risk established. A general procedure for managing risk is shown in Figure 6.1. This may

FIGURE 6.1

Risk management process

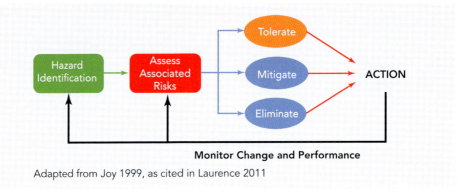

Adapted from Joy 1999, as cited in Laurence 2011

appear trivial and perhaps obvious, but there are some critical aspects and decision points in the diagram.

The first critical step is hazard identification. The presence of chemicals, high voltages, unstable rock, and the possibility of machine failure are examples of the many hazards in mining and mineral processing. Although guidance from manufacturers and regulatory bodies is available, there are no standard methods for hazard identification. There are many site-dependent factors.

The next step is to assess the possible risks associated with these hazards and establish a risk register, which is a full-time job for an engineer or other person trained in safety management. Again, there is no standard procedure for this, but experience in similar mines and guidance from manufacturers are used to assist the assessment.

Following a risk assessment, the decision to tolerate, mitigate, or eliminate must be made. Toleration of a risk depends only on the nature of the risk. Tolerance of a life-threatening risk or one that could lead to severe injury should never occur, although unfortunately, it does in some countries. In some cases, legislation may require mitigation or elimination of the risk combined with regular reporting to a regulatory body.

In other cases, a cost–benefit analysis might be done to decide whether to mitigate or eliminate the risk. This is usually framed as a comparison between the cost of minimizing or avoiding the damage (i.e., the benefit) and the cost incurred by the damage (Dhillon 2010). Equipment or system redesign for safer operation is one example. Hopkins (2015) questions this approach to risk management and suggests that accidents are the result of failures in control systems or procedures and it is the functioning of these systems and procedures that need attention.

The next two steps involve actually carrying out the decision made on what to do about the risks and monitoring the performance of the results. This requires strong, clear management statements. Monitoring the actions taken and their efficacy is an ongoing activity that provides feedback and informs any further action.

Safety Culture

The human factor continues to play a significant role in safety performance of a mine operation and in accidents. Several lines of research and surveys have revealed that, although eliminating or mitigating hazards may be successful at limiting the associated risks, people still behave unsafely and override any engineering design or controls (Kaila 2008; Simpson et al. 2009). Essentially, one cannot rely on "rule following" behavior of employees.

For this reason, continuous training, hiring practices, and demonstration of competency in safety practices at all levels have become key components of a mine operation. The goal is to induce habits, thinking, and behaviors that effectively become a culture based on safety. In modern mines, the process works such that no employee in any part of a mine operation is unaware of safety issues. Visitors to a mine feel the difference from the beginning when they are given a safety induction and then see that most actions and movements during the visit are deliberate and controlled, quite unlike what occurs in daily life outside the mine.

Industrial psychology and organizational behavior inform the process of establishing a safety culture and help to understand how it works in general and in particular. For example, Cooper and Phillips (2004) found that employee perceptions of the importance of safety training were predictive of actual levels of safety behavior. Also, in some mines, there may be large age differences among employees or diverse cultural and religious backgrounds, all leading to conceptions of safety that may differ from those of the desired safety culture. There is also evidence that employee literacy levels affect the comprehension of safety-related information, especially in a technology-rich environment (Guzman 2010). These situations have to be understood first and then managed continuously.

Two examples given next illustrate the range of possible safety cultures in mining operations. The first is an incident at an underground potash mine in Saskatchewan, Canada, in February 2014 in which some equipment caught fire and forced 96 workers to seek refuge in shelters for several hours until the fire was put out and the smoke cleared. No injuries or fatalities resulted. What is interesting is the attitude of the workers, some of whom thought it was a practice drill (Franner 2014):

> We regularly practice drills for these types of incidents, so when this happened, the workers just assumed it was another drill. There

was no panic. Everybody just did what they were supposed to do,
which is to go to a shelter area or to create a dead-end zone area
where there is no air circulation.

At the opposite end of the spectrum is the August 2010 rock collapse at the San
José copper-gold underground mine in Chile. The collapse trapped 33 miners for
69 days before they were rescued, one by one, using a capsule lowered through
a wide-diameter hole drilled to a refuge where the miners had managed to find
shelter. Before the collapse, there were signs of rock instability in the mine, but
these signs were apparently ignored. There were also numerous safety infractions,
such as missing ladders in the ventilation shaft that might have allowed the miners
to escape, but the mine was allowed to continue operation. This is indicative of an
absent safety culture (Pallardy 2023).

Despite Everything

Accidents and disasters still happen. The ultimate solution is to remove humans
from the mining system altogether. This will naturally happen if mining is done
from the deep ocean floor or in deep (more than 1,000 m) underground mines
using remotely operated machines. The question is whether remotely operated or
autonomous technologies can be developed to do mining efficiently at any depth
or, more generally, in any situation.

HEALTH

The physical and mental health of mine workers and of people in communities
near mines is a major concern of mine developers and managers. Issues relating to
personal health, health while working (occupational health), and community health
need to be considered when developing or operating a mine.

Personal Health

The diet, level of exercise, amount of sleep, stress level, and possible substance
abuse of a worker have a direct effect on safety at the mine. In addition, workers
can be exposed to all kinds of diseases and infections at the mine site and in the
community in which they live. Travel to and from the mine site and the physical
work environment (e.g., heat, humidity, cold) can all affect worker health. Regular
assessment of worker health as well as reviews of health risks are common practices
at modern mine operations.

Occupational Health

Occupational health is an interdisciplinary subject involving at the least medicine,
psychology, and ergonomics. It aims to promote the physical, mental, and social
well-being of workers by prevention of or protection from harmful situations.

Engineering plays a role in the design of systems that prevent and protect harmful situations in mine work environments. Monitoring and measurement of these systems is done to ensure performance. It is also necessary for employees to be informed and involved in the design and operation of the work environment and the machines within it—this is part of the safety culture.

Community Health

Some or all of the employees at a mine may come from the surrounding community; therefore, health in the community is just as important as the personal and occupational health of workers at a mine. In addition, community health is a "two-way street" in that while there may be health problems in the community, the mine emits solid and liquid wastes that can affect community health.

Investment in health-care systems may also be done if the site is remote or in a region where there is little community health care. In developing countries, it is common to see small hospitals and other infrastructure (e.g., water supply and sewage systems) built near an operation by a mining company as part of a corporate social responsibility program. There is a strong business case for this level of investment in community health and infrastructure (Humphreys 2000).

Mine wastes can have significant health effects on workers and the local population although modern, well-designed and well-operated mining and processing methods will prevent or greatly reduce the potential effects (Plumlee and Morman 2011). Mine waste emissions that must be considered in the design of a mine waste management scheme are:

- Solid wastes such as tailings, waste rock, and slag
- Liquid wastes from spent reagents or chemicals, and metal-contaminated water such as acid mine drainage
- Fumes and particulate emissions such as dust that can be inhaled or ingested into the respiratory system

Although spent reagents and other chemicals used in mineral processing can be harmful to humans, exposure to such substances is easily prevented. Other types of mine waste, such as tailings and waste rock, are not inherently harmful. The problem is that physical and chemical processes act on these wastes to generate sufficient concentrations of substances that could be harmful to humans. The pathways for these substances from source to ingestion may be complex and indirect; for example, from ingestion of a waterborne substance into fish and then into humans who eat the fish. Prevention of the physical and chemical processes that generate potentially harmful substances is a key focus of mine waste management.

DOSE AND TOXICITY

Dose is the intensity and duration of the exposure to a substance. It is a key factor in determining potential health effects. *Toxicity* is the degree to which the substance can cause damage to any organism, including humans. Toxicity is dose dependent. Dose and toxicity are analogous to hazard and risk—just as risk *is* dependent on a hazard, toxicity *is* dependent on the dose.

The term *toxic* is often used incorrectly when describing mine waste or any other kind of industrial waste. This is because it sounds really bad and grabs attention. However, whether something is toxic depends on the dose, not on the substance.

For example, pure water in sufficiently large doses can be toxic and potentially fatal to humans because a very large amount could cause low electrolyte levels in the body, which disturbs brain function.

Mining and processing practices in developing countries, particularly those of small-scale operators, are a community health concern; there are many examples of efforts by international health organizations and mining companies operating in such countries to improve practices. However, the fundamental problem is knowledge transfer and education of small-scale mine operators. Evans (2015) and Veiga et al. (2015) give examples of such educational initiatives.

MINING AND SOCIETY

Mining operations exert inevitable economic and social effects on local communities. Even in a community quite distant from a mine, the social fabric will be affected if mine employees come from the local community, perhaps in a fly-in, fly-out arrangement.

But mining has effects beyond a local community and its economy. For example, often overlooked and poorly understood is the fact that geology has placed large resources of the metals that society needs in countries that have limited capacity for regulation of the effects of large industrial operations on their environment, economy, and society. The result is an inequitable global distribution of the environmental, economic, and social impacts of mining.

Any mining company, regardless of its size, must take account of its place and role in a community and in society in general. The topic of the relationship between mining and communities and society is vast and it is evolving under the influence of developments both within and outside the mining industry.

Globalization

Globalization is the growth of international trade in goods, services, capital, and labor that results in an integrated global market in which all types of companies operate. Communication technologies and the internet have effectively eliminated borders in this market.

The size and market influence of corporations give rise to the perception that globalization is being driven by the private sector and not governments. Mining companies and suppliers are certainly large enough to have significant influence in a global market. In many ways, they can be agents of globalization. For example, an equipment supplier based in North America or Europe might supply machines to several mines located almost anywhere in the world. A regional subsidiary of the supplier might be established in one country to provide maintenance services to mines in the region. Replacement parts would be shipped from a global network of manufacturing plants. A local or expatriate workforce could operate the equipment.

The metal supply system discussed in Chapter 7 is truly global because minerals are extracted from mines in a number of countries, metals are produced in a number of different countries, and buyers of the metals may be almost anywhere. A massive global logistics "machine" connects all of the parts of the system.

Stakeholders

A mining company can be thought of as a set of inputs from investors, suppliers, and employees each contributing to an output—one or more mineral products. Each contributor expects to be compensated based on what a competitive market will offer. An alternative viewpoint is that the company has *stakeholders*, defined as any persons or organization who could be harmed by or benefit from the operation of the company, and who therefore have an interest in the operation of the company. All stakeholders have influence and some make a direct or indirect material contribution to the company operation.

The list of potential stakeholders is long and includes internal stakeholders, such as shareholders and employees, and many external stakeholders such as communities, governments, partner companies, suppliers, labor unions, professional organizations, and non-governmental organizations (or civil society organizations).

Owing to globalization and the connectivity provided by the internet, a mining company might find itself operating within a *network of stakeholders*, where the nature of relationships will differ depending on the project and its location. A resulting business activity is that all stakeholders will require variable amounts of company attention and resources at different times.

Corporate Social Responsibility

In addition to its physical and environmental "footprint" (see "The Metal Supply Footprint" section), the mine also has an economic and social footprint, which means that the company that owns and operates the mine must consider the potential effects of its actions. In other words, the company has a *corporate social responsibility* (CSR) that makes it accountable to its stakeholders, itself, its reputation, and the values it espouses.

Externally, a mining company is really no different from a company in another industry—it has stakeholders and a CSR. However, the complexity of stakeholder networks and relationships increases with the number and geographic extent of its projects. The relationships must be monitored and managed in order for the company to implement its CSR and for it to continue its operations. Employees of some mining companies are entirely devoted to CSR activities and to maintaining and managing the associated relationships. Unlike some other aspects of corporate governance, for example, the need for an audit committee, the scope and type of CSR activities are entirely voluntary.

Bowen et al. (2010) defined a continuum of CSR activities that can apply to any industry. These are illustrated in Figure 6.2. Fraser (2017) provided examples in mining contexts. On the left is transactional CSR. This is typically investments made by a mining company in infrastructure, buildings, or other initiatives that are of value to a local community, but do not contribute to the mining operation. They might be purely philanthropic in nature. The value of these kinds of investments lies in creating awareness of the company if it is new to the region and in building the trust needed to establish a more collaborative relationship in the future. However, conflict can result if different members of the community want different investments. There is also the potential that the investment is viewed as compensation for changes in land or water use caused by the mining operation; note that other forms of CSR involve investing more time and money in mitigation of these changes.

FIGURE 6.2

The continuum of corporate social responsibility

Transitional CSR is where the investments benefit both the company and the community. An example is training programs that enable members of the local community to work at the mine, thus providing career opportunities for community members and an accessible labor force for the mine. Local procurement of goods and services is another example. The lingering question about these kinds of CSR investments is "What happens when the mine closes?" Successful implementation and management of transitional CSR requires long-term thinking and planning.

A transformative CSR initiative is also designed to result in benefits for both the company and the local community, the difference being that the initiative is strategic and part of the core business of the mining company and its operation. Transformative CSR is best exemplified by shared value initiatives in which the business/strategic interests of the company are aligned with community needs. The results of *creating shared value* (CSV) are benefits to the mine during its operation and often lasting benefits to the community. Porter and Kramer (2006, 2011) introduced the concept of CSV as a possible application in any industry.

Fraser (2017) describes an example of CSV at the Cerro Verde mine in Peru where a public–private partnership between the community of Arequipa and the mine resulted in solutions to problems of water supply and quality that provided benefits for both the community and the mine operation. Another example of CSV in the mining industry is the eMalahleni water reclamation plant in South Africa that treats underground mine water at three coal mines operated by Anglo American (UNFCC 2012). The treated water is used in coal processing and, following treatment, as a water supply for local communities. Gypsum obtained as a result of water treatment is used to make construction materials. Further examples of CSV in mining can be found in Hidalgo et al. (2014).

Bowen et al. (2010) noted that the most common form of CSR is transactional because it involves mostly one-way communication from company to community. Transformative CSR is rare mainly because it requires organization, planning, negotiation, as well as community and company resources. This continues to be the case in the mining industry.

A company in the mining industry operates in a competitive and risky environment. The competition is to have access to high-quality resources at the lowest possible cost. There are many risks, but the risk of project interruption is the most severe to the continued operation or even the existence of the company. CSR initiatives can provide access to resources, reduce or eliminate the costs of disruption, and mitigate risk over the life of an operation. However, this should not be the goal of CSR initiatives. Many more benefits to the company and to stakeholders are possible if the social, economic, and environmental activities of CSR initiatives are aligned with business goals and strategy (O'Faircheallaigh 2010; Frederiksen 2018).

Social License to Operate

During the 1980s, there was a significant global increase in mining development, mostly by multinational mining companies investing in less-developed countries. Communities in those countries found that the practices of some mining operations could impose significant hardship or compromise the survival of their community, but existing regulations were unable to enforce less harmful practices. Consequently, communities demanded a greater role in decision-making. Without such participation, mining operations were exposed to significant risks of project interruption or cancellation.

Mining companies had to obtain ongoing approval for projects from the community and other stakeholders. Such approval is called the *social license to operate* (SLO), a term coined in 1997 by James Cooney who was an executive with Placer Dome at the time. As with any concept involving interaction with society, SLO required empirical observations of actual practice before becoming better understood. Academic and industry studies on SLO began to appear in the early part of this century (Joyce and Thomson 2000; Nelsen 2006; Thomson and Boutilier 2011).

An SLO must be earned. It cannot be obtained by a specific sequence of actions, and it cannot be bought by different forms of compensation. Prno (2013) used four case studies located in North and South America and Asia to understand common features of earning an SLO. The dynamics of an SLO are illustrated by Boutilier and Thompson (2018) who describe a case where a mine in Bolivia earned an SLO, lost it, and then regained it. An SLO must be maintained after it is earned.

SLO is also used as a risk mitigation strategy, a means to obtain continued access to mineral resources, but detached from the actual needs of stakeholders (Owen and Kemp 2013). What is needed are more concrete indications of SLO and, as with CSR, meaningful inclusion of stakeholders in company strategy.

CSR activities might provide such indications because they are usually related to visible items or actions. Transactional CSR, such as a philanthropic donation, might be an indication of an SLO, but this would depend on the level of understanding the community has about a proposed mine. A stronger argument for existence of an SLO can be made if a form of transformative CSR, such as a company–community collaboration, is present. But, as pointed out earlier, this type of CSR is rare.

U.N. Declaration on the Rights of Indigenous Peoples

The United Nations Declaration on the Rights of Indigenous Peoples (UNDRIP) is an agreement on the rights of Indigenous peoples around the world. It was adopted by the U.N. General Assembly in 2007 (UNGA 2007). UNDRIP is a comprehensive agreement that deals with human rights issues, such as cultural rights and identity, and the rights to education, health, employment, and language. Based on this, Indigenous people are considered *rightsholders*, not stakeholders. UNDRIP is not a response to the economic or social effects of mining (or to any other form of economic development).

UNDRIP is not legally binding, but it does represent a commitment by its adopters to follow a set of principles when interacting with Indigenous people. However, a clause in UNDRIP gives Indigenous people the right to self-determination, which can be interpreted as statehood. This led to legal and constitutional barriers to its legal adoption in some countries. This was the reason Australia, Canada, New Zealand, and the United States voted against UNDRIP in 2007. However, by 2010, each of the four countries adopted it. Flanagan (2020) describes the case of Canada:

> The Declaration calls for "free, prior, and informed consent" (FPIC) by Indigenous peoples before projects such as mines can be considered or developed on lands to which they may have a claim. Free means free from force, Prior means before any project decisions are made, and Informed means well-informed by the project developer and independent experts.

The need for FPIC results from adoption of UNDRIP by a government and therefore it must be achieved before a government will approve a mining project. However, FPIC does not imply that an SLO has been obtained. An SLO does not involve the government and it must be maintained during the life of the project.

Indigenous Mining Agreement

An Indigenous mining agreement (IMA) is a privately negotiated, legally enforceable contract between a mining company and an Indigenous community. It is known by different names depending on the jurisdiction or the intent. Table 6.1 contains a list of common names for IMAs.

An IMA addresses how a mining project will be developed and managed to reduce or mitigate adverse impacts of the project on the community and its cultural heritage and to provide benefits to the community. (Other projects, such

TABLE 6.1

Types of Indigenous mining agreements

Voluntary agreements	Shared responsibility agreements
Community development agreement	Investment agreements
Partnership agreements	Exploration agreements
Participation agreements	Benefits sharing agreements
Impact benefit agreement	Social trust funds
Community contracts	Empowerment agreements
Landowner agreements	Community joint ventures

Adapted from World Bank 2012

as hydroelectric schemes and pipelines, can also lead to IMAs.) Reduction or mitigation of adverse impacts typically involve measures that exceed regulatory requirements. Benefits can include regular monetary payments, royalty agreements depending on production, employment opportunities, training, preferential contracts, support of local businesses, and development of infrastructure.

O'Faircheallaigh (2016) provides examples of impact benefit agreements (IBAs) in Australia and Canada. In Canada as of 2023, there were 434 active IMAs, 75 of which were IBAs (NRC 2020). IMAs have become commonplace because public regulation of projects in mining jurisdictions was unable to adapt to conditions at different mining projects, making it difficult and costly to minimize the effects of projects on communities. However, the ability to insert project-specific provisions into an IMA provides the necessary flexibility to address local conditions (O'Faircheallaigh 2015). A large body of knowledge and expertise in negotiation of IMAs has developed (Gibson and O'Faircheallaigh 2015; Gunton et al. 2020).

An IMA is not compulsory, but in some Canadian and Australian jurisdictions, it is required before a mining permit will be issued. Although negotiating an IMA may be part of a process to obtain an SLO, it does not constitute an SLO because its intent is compensation and mitigation.

MINING AND THE ENVIRONMENT

The scope of mining and the environment as a subject is vast and interesting because it involves many different types of engineering and science. In the following, two topics are discussed, mainly to give an idea of the issues and challenges associated with the supply of metals and the environment. An entire book has been written on the subject (Spitz and Trudinger 2019).

The Metal Supply Footprint

The metal supply industry, which includes mining, causes different impacts. The preceding section concerned its impacts on society. The following describes its impacts on the environment in terms of the totality of its effects, or *footprint*.

There is the obvious physical footprint of open pit mining, waste dumps, and tailings storage facilities. Satellite imagery has been used to measure the global physical footprint of mining. Estimates vary between 25 and 50 million km^2 depending on the boundaries assumed for mining operations (Sonter et al. 2020; Tang and Werner 2023). The upper limit is comparable to the physical footprint of agriculture (Ritchie et al. 2022).

The metal supply industry also leaves a footprint on biodiversity, which is the variety of life in an area including animals, plants, and microorganisms. An important aspect of biodiversity is the complex relationships between these lifeforms and their habitat. Many of these relationships involve biochemical processes that provide important benefits to humans, including agricultural land for crop growth, clean air, and medicines. The metal supply industry affects biodiversity at different spatial scales and can lead to loss of biodiversity. For example, land clearing for an open pit mine, access roads, and other infrastructure causes disruption of animal habitat. Air emissions from a smelter may be harmful to animals and plants over a very large area. Sonter et al. (2020) advocate for long-term strategic assessment and planning in collaboration with mining companies to improve biodiversity outcomes at all scales.

Contaminant discharge from a mining operation into streams, lakes, or oceans is another part of the mining footprint. Fortunately, most of these discharges are well understood and can be controlled or prevented. Acid mine drainage is one example. However, some types of discharge are difficult to eliminate or control. Heavy metal contamination caused by artisanal gold mining is a serious problem that could be eliminated if artisanal miners could be encouraged to use alternative methods for gold extraction (Veiga et al. 2015). Control of selenium contamination from metal sulfide and coal mining is a challenge, but advances in its control have been made (Devi et al. 2021; Ostovar et al. 2022).

Metal Supply and Climate Change

Climate change refers to the increase in atmospheric temperature since 1900. The increase is caused by the emission of greenhouse gases (GHGs) due to the use of fossil fuels. GHGs, such as carbon dioxide and methane, trap the heat radiated from Earth in the atmosphere causing the temperature to rise. Compared to the 1951–1980 average, the mean global atmospheric temperature has increased by 1°C (NASA, n.d.).

Metal production involves a "value chain" consisting of mining, processing, refining, and transport of mineral products, each of which requires energy and is therefore a potential source of GHG emissions. There are three types of GHG emissions that depend on the stage of production in the value chain. These apply to any industry but for a mining operation, they are:

- Scope 1—emissions are GHG emissions from operating assets of the mine, such as haul trucks or diesel generators.
- Scope 2—emissions are due to power purchased by the mine and generated off-site by a method that emits GHGs.
- Scope 3—emissions occur beyond the mining operation and occur upstream or downstream in the value chain.

Estimates of Scope 1 and 2 emissions by the mining industry are relatively small, ranging between 4% and 10% of global GHG emissions (Nuss and Eckelman 2014; Delevingne et al. 2020; Azadi et al. 2020). For comparison, energy use in buildings is 17.5% of global emissions while agriculture, forestry, and land use account for 18.4% of global emissions (Ritchie 2020; Ritchie et al. 2020).

Efforts to reduce the Scope 1 and 2 emissions due to mining are focused on improving efficiencies of processing, particularly crushing and grinding; on energy recovery (Baidya et al. 2020); and on the use of renewable energy systems to supply electricity for transportation and for thermal energy needs (Igogo et al. 2021).

Scope 3 emissions include the upstream emissions generated in the manufacture of mining or processing equipment as well as the downstream emissions caused by metal or mineral product use in manufacturing. For example, power plants that use thermal coal are responsible for 20% of global GHG emissions (Birol and Malpass 2021). In steelmaking, the chemistry of steel production from iron ore is responsible for 7% of global GHG emissions while the electricity used is responsible for as much as another 2%–3%, depending on the type of power generated (Wright et al. 2023).

Should a mining company be concerned about Scope 3 emissions and assume some responsibility for them? There is a moral argument for doing so. But there is also a strong economic argument, which has been convincingly presented by Rohitesh Dhawan, president and CEO of the International Council on Mining and Metals (Dhawan 2022).

Many investors want minimum to zero exposure to companies that directly or indirectly contribute to GHG emissions, and many consumers favor products that contain metals whose production has not contributed to GHG emissions, that is, ones that have a low- or zero-carbon footprint. Thus, to attract investment and maintain a market for metals and manufactured products, it is worth knowing the amounts of Scope 3 GHG emissions in the entire value chain and to make contributions to their reduction.

In an effort to avoid or minimize Scope 3 emissions, mining companies have entered into partnerships with steelmaking operations to investigate ways to reduce emissions. One promising alternative is to replace the metallurgical coal used to reduce the iron ore with hydrogen made by electrolyzing water using electricity produced by renewable energy systems (e.g., Hybrit fossil-free steel). The result would be steel produced with no fossil fuel input.

The complexity of calculating the amount of Scope 3 emissions is well illustrated by two annual GHG emission reports (BHP 2023; Rio Tinto 2022). The International Council on Mining and Metals has published accounting and reporting guidance for Scope 3 emissions (ICMM 2023).

MINING AND SUSTAINABILITY

Sustainability is the result of meeting the current needs and demands of society without compromising the needs and demands of future generations. Environmental, social, and economic needs and demands are considered together, not individually. This suggests practices and processes that can be maintained over a long period. Such practices and processes constitute what is known as *sustainable development* (WCED 1987).

Embedded in sustainability is the concept of intergenerational equity—nothing done today should hinder future generations. The natural question is: "Will the metal supply footprint create a legacy that could hinder future generations?" A complete answer involves forecasting the future needs and demands of society. CSR and shared value initiatives in mining make it possible to provide long-lasting benefits to a community and to society, and, in this sense, mining can be a force for sustainable development (Fraser 2017, 2019). However, is this enough? Parts of the metal supply footprint may still be present. For this reason, current metal supply activity and research needs to find ways to minimize the footprint.

One clearly unsustainable aspect of mining is resource depletion—extraction activities today will leave fewer resources available for future generations. However, this assumes that future generations will obtain their materials from the same sources and in the same way we do today. Chapter 8 makes a case for a scenario where future sources of materials and methods for their extraction will be different from those of today.

Sustainable Development Goals

In 2015, all member states of the United Nations adopted the 2030 Agenda for Sustainable Development (UNGA 2015). The core of this agenda is 17 sustainable development goals (SDGs), which are ambitious targets to be achieved by 2030 that would end poverty, improve health and education, reduce inequality, spur economic growth, deal with climate change, and preserve biodiversity. Partnerships between nations, industries, and groups of companies are envisioned as a leading mechanism to achieve these goals.

The goals are shown in Figure 6.3. Of particular interest is that a sustainable supply of metals is not mentioned in any of the SDGs, yet such a supply is and would be essential to achieving the SDGs (Franks et al. 2022). The difficulty with including metals in the SDGs may be that, although mining can be a force for sustainable development, the current metal supply paradigm and footprint is often difficult to reconcile with or connect to some of the SDGs.

FIGURE 6.3

The U.N. Sustainable Development Goals

Source: United Nations Sustainable Development Goals. Copyright © United Nations.*

* The content of this publication has not been approved by the United Nations and does not reflect the views of the United Nations or its officials or Member States. www.un.org/sustainabledevelopment.

Fraser (2017, 2019) explored the possibilities for contributing to the U.N. SDGs through shared value initiatives at or near mining operations. Implementation of some of the technologies described in Chapter 8 would contribute to advancing SDGs such as SDG 7–9, 12, and 13.

Environmental Stewardship, Social Responsibility, and Corporate Governance

ESG stands for *E*nvironmental stewardship, *S*ocial responsibility, and corporate *G*overnance. Since the early part of the 21st century, these have become significant considerations for investors in companies, including mining companies. Previously, the only relevant indicator of company performance was shareholder return, the so-called Friedman doctrine (Friedman 1970). During the 1990s, an alternative to shareholder return arose, exemplified by the Triple Bottom Line (Elkington 1999) where environmental quality and social justice were included with profitability as measures of the value of a company. This and other events gave rise to a new investment style sometimes called *social impact investing*. Its aim is to encourage corporations to include environmental and social issues in their decision-making.

Company management of ESG is rated by one of several ESG rating agencies (e.g., Sustainalytics, n.d.; LSEG, n.d.). Ratings are measures of ESG risks that have financial consequences. They are based on assessments of factors such as carbon footprint, health and safety, water usage, community development efforts, and employee diversity. Investors use these ratings to design investment portfolios. A company can use the ratings from one agency to measure performance against its peers. However, the ratings of one company by different agencies can differ considerably.

Mining companies, and individual operations within a mining company, can make voluntary disclosures to industry-sponsored programs that have established standards for responsible mining (IRMA, n.d.) or for sustainable mining (TSM 2023). These programs have developed in response to consumer demand for knowledge about the sources of metals in products, such as their carbon, energy, or water footprint, or the existence of human rights violations.

Much importance has been attached to ESG ratings, mainly because the issues associated with ESG are a genuine concern to investors and other stakeholders. However, assessments of E, S, and G are non-financial and will depend on the way a company discloses its ESG performance. As a result, the metrics contain uncertainty and subjective elements (Ilango 2022).

Long ago, Lord Kelvin (1824–1907) and Peter Drucker (1909–2005) made statements about the importance of measurable numbers in the description or management of something. These statements still apply.

The Equator Principles

The Equator Principles (Equator Principles Association 2023) are a set of standards developed in 2003 by the International Finance Corporation (IFC), a member of the World Bank group, for use by financial institutions to assess environmental and social risks of projects. Credit risk can result if environmental and social risks are not managed to a minimum standard. Several financial institutions have committed to not provide loans to mining or other projects that do not comply with these standards. Currently 140 financial institutions in 39 countries adhere to the Equator Principles.

Artisanal Mining

Artisanal mining, abbreviated ASM, touches on every technical topic in this book. It is a significant part of mining, and therefore the best location for this section became a question. However, because ASM provides income for about 45 million workers in 80 countries (World Bank 2020; Artisanal Mining, n.d.), and because this chapter is where people and interactions between mining and people are discussed, this chapter is the right location.

ASM is the extraction and processing of ore from rock or from sediments. Manual methods or simple machinery are used for mining. Gold in alluvial deposits is the most common target for artisanal miners. Other targets include diamonds; colored and semi-precious stones; the metals cobalt, tantalum, lithium, tin, and tungsten; and industrial minerals such as sand and mica. Mining for these other targets usually requires construction of a mineshaft (by hand) and working underground in dangerous, unsupported openings.

ASM is an inefficient method of metal production, but that is the least of its problems. Environmental damage, contamination, and health problems result from all types of ASM. Artisanal gold mining provides a good illustration. Excavation of gold-bearing alluvial sediments alters drainage patterns in rivers and on land. Manual panning of the excavated sediments is done to separate the lighter sediments from the heavier gold particles, as shown in Figure 6.4a. Manual excavation and panning can result in over-exertion, accidents, or heat stress. Mercury is mixed with the sediments remaining after panning to form an amalgam with the free gold particles. The mercury is boiled off, as shown in Figure 6.4b, leaving the

FIGURE 6.4

Artisanal mining

(a) Excavation and panning of sediments
(Courtesy of Reuters/Luc Gnango)

(b) Boiling off mercury from gold-mercury amalgam
(Courtesy of Artisanal Gold Council)

gold behind. Mercury contamination of soils and surface water runoff results and mercury poisoning caused by mercury vapor inhalation is highly likely.

Gold in rock extracted from an artisanal underground mine is often contained within sulfide minerals. The rock is subjected to grinding to liberate the gold particles from the sulfides, and cyanide is added to dissolve the gold. Zinc is used to precipitate the gold. (See Chapter 3.) All this is done in containers or tanks, but control is required to prevent leaks or spills.

ASM typically has no legal or regulatory framework and no access to a formal metals market. It is a breeding ground for many types of criminal activity. Proceeds of ASM have been used by governments and by rebel groups to fund civil wars. The proceeds have also been used by organized crime to fund illicit drug manufacture and smuggling. Crime proceeds can be used to buy mineral products, especially gold, from miners, thus laundering the money (Naré et al. 2022). Both these situations provide opportunities for forced labor, child labor, extortion, human rights abuses, and so forth.

However, ASM supports a large number of people who have no other source of income; that is, it is subsistence mining. Annual gold production by ASM is typically about 20% of total gold production, which was 3,612 t (metric tons) in 2022 (World Gold Council 2022a). This is a significant amount of potential revenue. If ASM were organized and managed properly (i.e., formalized), it could contribute to achieving SDG 8: Decent Work and Economic Growth. There have been several initiatives to improve ASM working conditions; encourage the use of more efficient, less contaminating methods; and reduce the associated risks (World Gold Council 2022b; Artisanal Gold Council, n.d.). A more recent development is collaboration of ASM miners with a mining company that has an operation in the same area. The World Gold Council (2022b) has described examples where the company supports training, adoption of safer mining practices, clean methods of processing, and provides access to global metals markets.

SUMMARY

This has been a whirlwind tour of a number of diverse and seemingly disparate topics. Metals and minerals are the "connective tissue" that tightly binds these topics together. But, more importantly, these topics connect the metals and minerals industry to other industries and to society. What is interesting is how fast these topics are adapting and changing.

Different forms of partnerships were seen to be a prominent mode of connection, for example, industry–academic partnerships for training the skilled professionals needed in the mining industry. Development and maintenance of a safety culture is best achieved by a partnership between employees and management. CSR for a mining company is always a partnership with a community and likely other stakeholders. Efforts to limit GHG emissions in order to reduce or eliminate the carbon footprint of products containing metals is done via partnerships between mining companies and manufacturers. In some locations, partnerships between ASM miners and mining companies have successfully reduced the environmental and social damage caused by ASM.

More of these partnerships are to be expected. It is difficult to think of other ways to achieve the desired goals.

REFERENCES

Abenov, T., Franklin-Hensler, M., Grabbert, T., et al. 2023. Has mining lost its luster? Why talent is moving elsewhere and how to bring them back. *Our Insights*, February 14. McKinsey & Company. www.mckinsey.com/industries/metals-and-mining/our -insights/has-mining-lost-its-luster-why-talent-is-moving-elsewhere-and-how-to-bring -them-back. Accessed August 2023.

Artisanal Gold Council. n.d. Mission and vision. https://artisanalgold.org/mission-vision/. Accessed October 2023.

Artisanal Mining. n.d. Map 1: Number of ASM miners per country. ASM Inventory: World Maps of Artisanal and Small-Scale Mining. http://artisanalmining.org /Inventory/. Accessed October 2023.

Azadi, M., Northey, S., Ali, S., et al. 2020. Transparency on greenhouse gas emissions from mining to enable climate change mitigation. *Nature Geoscience* 13:100–104.

Baidya, D., Rodrigues de Brito, M.A., and Ghoreishi-Madiseh, S.A. 2020. Techno-economic feasibility investigation of incorporating energy storage with an exhaust heat recovery system for underground mines in cold climatic regions. *Applied Energy* 273:115289.

Banta, J.L., Barton, I., and Hutson, L. 2021. Where have all the mining engineering students gone? *Mining Engineering* (February): 25–28.

Barton, I., Banta, J.L., and Hutson, L. 2021. How to get more students to major in mining engineering? *Mining Engineering* (February): 30–34.

BHP. 2023. *Scopes 1, 2 and 3 GHG Emissions Calculation Methodology*. www.bhp.com /-/media/documents/investors/annual-reports/2023/220822_bhpscopes12and3 emissionscalculationmethodology2023.pdf. Accessed August 2023.

Birol, F., and Malpass, D. 2021. It's critical to tackle coal emissions. International Energy Agency. October 8. www.iea.org/commentaries/it-s-critical-to-tackle-coal-emissions. Accessed November 2023.

Boutilier, R.G., and Thompson, I. 2018. *The Social License: The Story of the San Cristobal Mine*. London: Routledge.

Bowen, F., Newenham-Kahindi, A., and Herremans, I. 2010. When suits meet roots: The antecedents and consequences of community engagement strategy. *Journal of Business Ethics* 95:297–318.

Cooper, M.D., and Phillips, R.A. 2004. Exploratory analysis of the safety climate and safety behavior relationship. *Journal of Safety Research* 35:497–512.

Delevingne, L., Glazener, W., Grégoir, L., et al. 2020. Climate risk and decarbonization: What every mining CEO needs to know. McKinsey & Company. www.mckinsey.com /capabilities/sustainability/our-insights/climate-risk-and-decarbonization-what-every -mining-ceo-needs-to-know. Accessed November 2023.

Denison, E., and Pringle, R. 2023. How a remote Australian mine became a gender -balanced workplace. *Harvard Business Review*. July 13; updated August 25. https://hbr .org/2023/07/how-a-remote-australian-mine-became-a-gender-balanced-workplace. Accessed November 2023.

Devi, P., Singh, P., Malakar, A., et al., eds. 2021. *Selenium Contamination in Water*. New York: Wiley-Blackwell.

Dhawan, R. 2022. Why mining is crucial for sustainability w/ Rohitesh Dhawan. In *Sustainability Decoded with Tim & Caitlin*. Produced by Persefoni and Hueman Group Median. Podcast, MP3 audio, 40:40. https://sites.libsyn.com/417785/10-why-mining -is-crucial-for-sustainability-w-rohitesh-dhawan. Accessed November 2023.

Dhillon, B.S. 2010. *Mine Safety: A Modern Approach*. London: Springer.

Elkington, J. 1999. *Cannibals with Forks: The Triple Bottom Line of 21st Century Business*. New York: Wiley.

Ellix, H., Farmer, K., Kowalik, L., et al. 2021. Why women are leaving the mining industry and what mining companies can do about it. *Our Insights*, September 13. McKinsey & Company. www.mckinsey.com/industries/metals-and-mining/our-insights/why -women-are-leaving-the-mining-industry-and-what-mining-companies-can-do -about-it. Accessed August 2023.

Equator Principles Association. 2023. The Equator Principles. https://equator-principles .com. Accessed November 2023.

Evans, R. 2015. Skills and education: The key to unlocking development from mining. *AUSIMM Bulletin*, February.

Fernandez-Stark, K., Couto, V., and Bamber, P. 2019. *Industry 4.0 in Developing Countries: The Mine of the Future and the Role of Women*. https://documents1.worldbank.org/ curated/pt/824061568089601224/Industry-4-0-in-Developing-Countries-The-Mine -of-the-Future-and-the-Role-of-Women.pdf. Accessed August 2023.

Flanagan, T. 2020. *Squaring the Circle: Adopting UNDRIP in Canada*. Vancouver, BC, Canada: Fraser Institute.

Franks, D.M., Keenan, J., and Hailu, D. 2022. Mineral security essential to achieving the Sustainable Development Goals. *Nature Sustainability* 6:21–27. doi:10.1038 /s41893-022-00967 9.

Franner, M. 2014. Underscoring the importance of safety: Mining industry seeks to be the best. *PotashWorks,* October 7. https://imii.ca/wp-content/uploads/2022/07 /Underscoring_the_importance_of_safety__Mining_industry_seeks_to_be_the _best___PotashWorks.pdf. Accessed November 2023.

Fraser, J. 2017. From social risk to shared purpose: Reframing mining's approach to corporate social responsibility. PhD dissertation, University of British Columbia, Vancouver, Canada. https://open.library.ubc.ca/collections/ubctheses/24 /items/1.0353179. Accessed November 2023.

Fraser, J. 2019. Creating shared value as a business strategy for mining to advance the United Nations Sustainable Development Goals. *The Extractive Industries and Society* 6:788–791.

Frederiksen, T. 2018. Corporate social responsibility, risk and development in the mining industry. *Resources Policy* 59:495–505.

Friedman, M. 1970. A Friedman doctrine—The social responsibility of business is to increase its profits. *New York Times*, September 13. www.nytimes.com/1970/09/13 /archives/a-friedman-doctrine-the-social-responsibility-of-business-is-to.html. Accessed November 2023.

Gibson, G., and O'Faircheallaigh, C. 2015. *IBA Community Toolkit, Negotiation and Implementation of Impact and Benefit Agreements.* Toronto, ON, Canada: Gordon Foundation. https://gordonfoundation.ca/. Accessed July 2023.

Gunton, C., Batson, J., Gunton, T., et al. 2020. *Impact Benefit Agreement Guidebook.* Burnaby, BC, Canada: School of Resource and Environmental Management, Simon Fraser University. http://rem-main.rem.sfu.ca/planning/IBA/IBA_Guidebook_2-24 .pdf. Accessed November 2023.

Guzman, M.-L. 2010. Study looks at literacy effect on workplace safety. *The Safety Magazine*, July 21. www.thesafetymag.com/ca/topics/psychological-safety/study-looks -at-literacy-effect-on-workplace-safety/186188. Accessed November 2023.

Hidalgo, C., Peterson, K., Smith, D., et al. 2014. *Extracting with Purpose Creating Shared Value in the Oil and Gas and Mining Sectors' Companies and Communities.* www.fsg.org /resource/extracting-purpose/. Accessed July 2023.

Hopkins, A. 2015. How much should be spent to prevent disaster? A critique of consequence times probability. *Journal of Pipeline Engineering* 14(2):69–78.

Humphreys, D. 2000. A business perspective on community relations in mining. *Resources Policy* 26:127–131.

ICMM (International Council on Mining and Metals). 2023. Scope 3 emissions accounting and reporting guidance. September 1. www.icmm.com/en-gb/guidance /environmental-stewardship/2023/scope-3-emissions-accounting-and-reporting. Accessed September 2023.

Igogo, T., Awuah-Offei, K., Newman, A., et al. 2021. Integrating renewable energy into mining operations: Opportunities, challenges, and enabling approaches. *Applied Energy* 300:117375.

Ilango, H.J. 2022. *Greater ESG Rating Consistency Could Encourage Sustainable Investments.* Lakewood, OH: Institute for Energy Economics and Financial Analysis. www.ieefa.org. Accessed September 2023.

IRMA (Initiative for Responsible Mining Assurance). n.d. Standard [for responsible mining]. https://responsiblemining.net/what-we-do/standard/. Accessed August 2023.

Joyce, S., and Thomson, I. 2000. Earning a social license to operate: Social acceptability and resource development in Latin America. *Canadian Mining and Metallurgical Bulletin* 93:49–53.

Kaila, H.L. 2008. *Behaviour-Based Safety in Organizations: A Practical Guide.* New Delhi: I.K. International.

Laurence, D. 2011. Mine safety. In *SME Mining Engineering Handbook*, 3rd ed. Edited by P. Darling. Englewood, CO: SME. p. 1559.

LSEG. n.d. Sustainability strategy. www.lseg.com/en/sustainability-strategy. Accessed August 2023.

MIHRC (Mining Industry Human Resources Council). 2023. Mining for the future: The clean economy transition needs more people. https://mihr.ca/wp-content /uploads/2023/08/MIHR-Info-Graphic-Poster-EN.pdf. Accessed November 2023.

Naré, C., Crawford, A., Gronwald, V., et al. 2022. *Illicit Financial Flows and Conflict in Artisanal and Small-Scale Gold Mining: Burkina Faso, Mali, and Niger.* Winnipeg, MB, Canada: International Institute for Sustainable Development.

NASA (National Aeronautics and Space Administration). n.d. Earth observatory: World of change: Global temperatures. https://earthobservatory.nasa.gov/world-of-change /global-temperatures. Accessed August 2023.

Nelsen, J.L. 2006. Social license to operate. *International Journal of Mining, Reclamation and Environment* 20(3):161–162.

NRC (Natural Resources Canada). 2020. Lands and minerals sector—Indigenous mining agreements. https://atlas.gc.ca/imaema/en/index.html. Accessed July 2023.

Nuss, P., and Eckelman, M.J. 2014. Life cycle assessment of metals: A scientific synthesis. *PLOS One* 9(7): e101298. doi:10.1371/journal.pone.0101298.

O'Faircheallaigh, C. 2010. CSR, the mining industry and Indigenous peoples in Australia and Canada: From cost and risk minimisation to value creation and sustainable development. In *Innovative CSR: From Risk Management to Value Creation.* Edited by C. Louche, S.O. Idowu, and W.L. Filho. New York: Routledge. pp. 398–418.

O'Faircheallaigh, C. 2015. Social equity and large mining projects: Voluntary industry initiatives, public regulation and community development agreements. *Journal of Business Ethics* 132:91–103.

O'Faircheallaigh, C. 2016. *Negotiations in the Indigenous World: Aboriginal Peoples and the Extractive Industry in Australia and Canada.* New York: Taylor and Francis.

Ostovar, M., Saberi, N., and Ghiassi, R. 2022. Selenium contamination in water; analytical and removal methods: A comprehensive review. *Separation Science and Technology* 57(15):2500–2520. doi:10.1080/01496395.2022.2074861.

Owen, J.R., and Kemp, D. 2013. Social licence and mining: A critical perspective. *Resources Policy* 38:29–35.

Pallardy, R. 2023. Chile mine rescue of 2010. *Encyclopedia Britannica.* www.britannica .com/event/Chile-mine-rescue-of-2010. Accessed November 2023.

Plumlee, G.S., and Morman, S.A. 2011. Mine wastes and human health. *Elements* 7:399–404.

Porter, M., and Kramer, M. 2006. Strategy and society: The link between competitive advantage and corporate social responsibility. *Harvard Business Review* 81(12):78–93.

Porter, M., and Kramer, M. 2011. Creating shared value. *Harvard Business Review* 89(1/2):62–77.

Prno, J. 2013. An analysis of factors leading to the establishment of a social licence to operate in the mining industry. *Resources Policy* 38(4):577–590.

Rio Tinto. 2022. *Scope 1, 2 and 3 Emissions Calculations Methodology 2022.* www.riotinto .com/en/sustainability/climate-change. Accessed August 2023.

Ritchie, H. 2020. Sector by sector: Where do global greenhouse gas emissions come from? Our World in Data, September 18. https://ourworldindata.org/ghg-emissions-by -sector. Accessed November 2023.

Ritchie, H., Roser M., and Rosado, P. 2020. CO_2 and greenhouse gas emissions. Our World in Data. https://ourworldindata.org/co2-and-greenhouse-gas-emissions. Accessed July 2023.

Ritchie, H., Rosado, P., and Roser M. 2022. Environmental impacts of food production. Our World in Data. https://ourworldindata.org/environmental-impacts-of-food. Accessed September 2023.

Simpson, G., Horberry, T., and Joy, J. 2009. *Understanding Human Error in Mine Safety.* Surrey, UK: Ashgate.

Sonter, L.J., Dade, M.C., Watson, J.E.M. et al. 2020. Renewable energy production will exacerbate mining threats to biodiversity. *Nature Communications* 11:4174.

Spitz, K., and Trudinger, J. 2019. *Mining and the Environment: From Ore to Metal*, 2nd ed. New York: Routledge.

Sustainalytics. n.d. Company ESG risk ratings. www.sustainalytics.com/esg-ratings. Accessed August 2023.

Tang, L., and Werner, T.T. 2023. Global mining footprint mapped from high-resolution satellite imagery. *Communications Earth & Environment* 4:134.

Thomson, I., and Boutilier, R.G. 2011. Social license to operate. In *SME Mining Engineering Handbook*, 3rd ed. Edited by P. Darling. Englewood, CO: SME. pp. 1779–1796.

TSM (Towards Sustainable Mining). 2023. Towards sustainable mining: TSM initiative. https://tsminitiative.com/. Accessed August 2023.

UNFCC (United Nations Framework Convention on Climate Change). 2012. eMalahleni: Water reclamation plant: South Africa. https://unfccc.int/climate-action/momentum -for-change/lighthouse-activities/emalahleni-water-reclamation-plant. Accessed November 2023.

UNGA (United Nations General Assembly). 2007. *Declaration on the Rights of Indigenous Peoples.* www.un.org/development/desa/indigenouspeoples/wp-content/uploads /sites/19/2018/11/UNDRIP_E_web.pdf. Accessed August 2023.

UNGA (United Nations General Assembly). 2015. *Transforming Our World: The 2030 Agenda for Sustainable Development.* https://sdgs.un.org/2030agenda. Accessed November 2023.

Veiga, M.M., Angeloci, G., Niquen, W., et al. 2015. Reducing mercury pollution by training Peruvian artisanal gold miners. *Journal of Cleaner Production* 94:268–277.

WCED (World Commission on Environment and Development). 1987. *Our Common Future.* Oxford, UK: Oxford University Press.

World Bank. 2012. *Mining Community Development Agreements: Source Book.* Washington, DC: World Bank. http://hdl.handle.net/10986/12641. Accessed August 2023.

World Bank. 2020. *2020 State of the Artisanal and Small-Scale Mining Sector.* Washington, DC: World Bank.

World Gold Council. 2022a. Gold demand trends full year 2022. www.gold.org/goldhub /research/gold-demand-trends/gold-demand-trends-full-year-2022/supply. Accessed October 2023.

World Gold Council. 2022b. *Lessons Learned on Managing the Interface Between Large-Scale and Artisanal and Small-Scale Gold Mining.* London: World Gold Council. www.gold .org/esg. Accessed October 2023.

Wright, L., Liu, X., Wu, I., et al. 2023. *Steel GHG Emissions Reporting Guidance.* Washington, DC: RMI. https://rmi.org/wp-content/uploads/2022/09/steel_emissions _reporting_guidance.pdf. Accessed November 2023.

Mining and Money

This chapter provides answers to a number of questions: What products are produced at mines and processing plants? What are by-products, co-products, and metal equivalents? How are metal prices determined? What are resources and reserves, and how do publicly traded mining companies report them?

Figure 7.1 shows some of the products of mines. Each of these products is a *commodity*; that is, there is little to no physical or chemical difference between these products and similar products produced elsewhere in the world. They are sold on the open market.

1-kg gold bars
(Courtesy of Agnico-Eagle Mines Limited)

Copper concentrate being loaded into ship
(Courtesy of Copper Mountain Mining Corporation)

Iron ore pellets
(Courtesy of Iron Ore Company of Canada)

Coal stacker forming stockpile at Westshore coal terminal, Delta, British Columbia, Canada
(Courtesy of Westshore Terminals)

FIGURE 7.1

Common mineral products

THE METAL SUPPLY SYSTEM

Figure 7.2 shows the collection of participants in the global metal supply system. The flow of minerals and metals is essentially a linear sequence from mining companies (MineCos) who produce a mineral concentrate that smelters and refineries (RefineCos) turn into pure metal(s) using metallurgical processes. The metals are sold to manufacturers or other consumers through organized metal exchanges, or through metal traders if there is a small number of buyers and sellers (a "thin" market).

FIGURE 7.2

The metal supply system

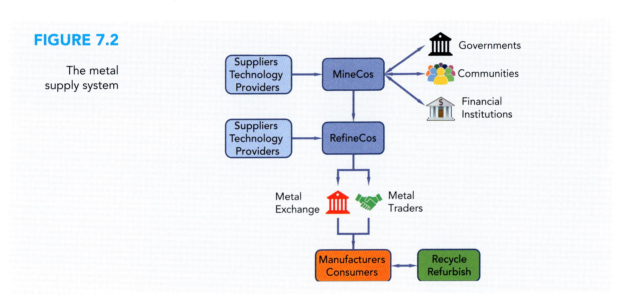

There is a diverse range of participants shown in Figure 7.2. Country governments are the owners of mineral rights, and mining companies pay some form of compensation (a royalty) to the government for the right to carry out mining in the country. Country governments also regulate mining practices. Communities are affected by mining operations and may depend on them for employment. Financial institutions supply loans for mine development and construction. Suppliers and technology providers, located inside or outside the country, may receive contracts to perform some of the activities in mining and metal production or to supply particular types of equipment. Manufacturers buy metals from exchanges or from metal traders. Independent companies may recycle end-of-life products to recover the metals in them or refurbish and return them to be used in a circular economy; otherwise, they become waste.

Another important aspect of the metal supply system not depicted by Figure 7.2 is its global scope. Geology and plate tectonics have placed large and rich resources of metals in developing countries that have limited to no capacity for development of industrial operations such as mines. However, mining and supplier companies in developed economies do have the capacity and are attracted to these large resource endowments. Consequently, they have become major participants in mining operations in developing economies, thus increasing the global scope of the industry—a form of globalization.

The global scope of metal supply is also strongly influenced by population growth and urbanization, both of which lead to demand for products or systems containing metals. World population was 5.3 billion in 1990, 8.0 billion in 2022 (World Bank 2023), and is expected to be 9.7 billion in 2050 (United Nations 2023). Urbanization was 43% of total population in 1990 and grew to 57% of total population in 2022 (World Bank 2023). The metal supply system will need to grow and adapt to satisfy the resulting demands.

At the end of the chapter is a list of books on the metals supply system and markets.

Mine Product Streams

There are five product streams depending on the type of ore:

1. In some rare cases, the ore from a mine is so rich that it does not require processing to make a concentrate. For some time, the ore from the Eskay Creek mine in British Columbia (Canada) was shipped directly to smelters and refineries in Quebec (Canada) or Japan.

2. Base metal sulfide ores are processed by flotation to produce a concentrate with a concentration of 20%–40% of a metal. Copper concentrates are typically 25%–30%. The concentrate is transported to a smelter/refinery complex where pure metal is made. Unless the mining company owns the smelter/refinery, the sale of concentrate is governed by a smelter contract. An alternative is to process the concentrate by pressure leaching and electrowinning to form pure metal.

3. Low-grade sulfide or oxide ores are leached in a heap leach pad to produce a leachate with low metal concentration. Next, the leachate is further concentrated by solvent extraction. An electric current is then applied to the resulting solution in an electrowinning cell and the metal is plated onto the cathode of the cell.

4. Precious metal ores are processed by leaching and electrowinning to produce an impure metal product that must be refined to produce pure (99.99%) metal. The impure metal at a gold mine is called *doré*, a mixture of 60%–90% gold and other metals, often silver. The impure metal is sold to an independent refinery under the terms of a refinery contract.

5. Iron ore, potash, and industrial minerals typically require some form of separation technology to produce a desired product. For example, flotation is used to obtain fertilizer-grade potassium chloride and to separate fine coal; grinders, cyclones, and magnetic separation methods are used to produce iron ore products. The processed ore is shipped to a buyer under terms of a delivery contract, which specifies the delivery times of required quantities and the required qualities or grades.

By-Products and Co-Products

More than one product is often produced by the same mine and mineral processing plant. Gold, silver, and molybdenum are found in copper ore bodies; lead and zinc are usually found together; and platinum is found in nickel deposits. The decision as to whether these metals are co-products (or joint products) or whether one or more of the metals is a by-product is important because it affects the allocation of costs.

The distinction between by-products and co-products is based entirely on the relative value of the quantities produced. The value of by-product production is low relative to the value of the primary metal produced. By-products are usually not important to the viability of the mine. Examples of by-products are the many metals produced from lead and zinc concentrates. In comparison, the relative values of co-products produced are similar, and they are usually important to the viability of the mine. Examples include lead and zinc, vanadium and uranium, and sometimes molybdenum in a copper mine.

By-products and co-products can be physically distinguished at a "split-off" point in the production process. The ease with which a split-off point can be identified will vary depending on the products, and there could be multiple split-offs depending on the number of products and the processes used to separate them. However, once the split-offs are identified, costs can be allocated in some rational manner. As shown in Figure 7.3, the costs of co-products can be allocated based on revenue or on some physical basis such as mass. The same approach can be used for by-products, but given that by-product production is relatively small, precise cost allocation is not required.

Current industry practice is to use the revenue from by-products as a deduction from the unit cost of production of the primary product. Thus, by-product gold credit for a copper concentrate might be used to reduce the operating cost of a copper mine. However, in some cases, the high prices received for a by-product cause

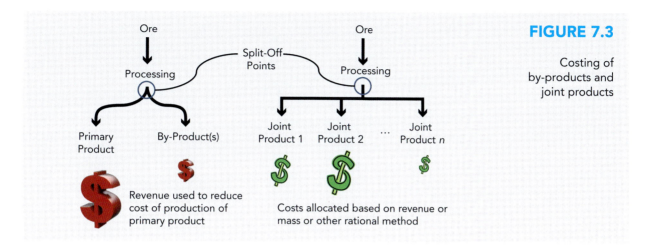

FIGURE 7.3

Costing of by-products and joint products

the deduction to be greater than the unit cost of primary metal production; that is, the net cost of production becomes negative. Varying quantities of by-product produced can also cause the cost of the primary product to vary considerably. The use of by-product credits can be misleading, and other measures of profitability must be considered.

Metal Equivalents

Metal equivalents express the quantity of metal in a polymetallic deposit in terms of the metal of most value (primary metal) in the deposit. The primary metal equivalent mass is given by

$$\text{primary metal equivalent mass} = \frac{\text{value or all recoverable metals in ore (\$)}}{\text{unit value of recoverable primary metal (\$/mass)}}$$

The equivalent grade is given by dividing the equivalent mass by the mass of the rock containing the metals. (Watch the units!) The value of the metals must include the recovery of the metals in the processing plant.

This calculation ignores grade variations within deposit and price (value) changes, which vary during the operation. The possible variation in these parameters can be significant, and the results become particularly confusing if different kinds of processing occur (e.g., recoveries differ in sulfide and oxide ore), or there are different grade zones within an ore body. Securities regulators (see the "Reporting Standards" section) allow reporting of metal equivalents, but require that the assumptions of grade, recovery, and price be explicitly stated. Easier, less prone to error, and more informative is to simply state the metal grades, tonnages of rock containing the grades, and estimated recoveries, from which the amount of recoverable metal can be computed. Given current prices, the total value can be computed.

METAL EQUIVALENT CALCULATION

Metal equivalents are similar to a weighted average. Consider a polymetallic deposit containing silver, lead, and zinc in a rock mass of 5 Mt (million metric tons). The concentrations, recoveries, and unit prices (later 2023) of these metals are as follows:

Metal	Grade	Recovery, %	Price, $US	Value, $US million
Silver	130 g/t	82	24.55/oz	420.7
Lead	1.52%	93	2,188.9/t	133.3
Zinc	2.15%	88	2,544.4/t	207.4
			Total value of all metals	761.4

Therefore, if silver is the primary metal, the equivalent silver mass is (1 troy oz = 31.1 g):

$$761.4/(24.55/31.1)/0.82 = 1,176.3 \text{ Mg silver}$$

Consequently, the equivalent silver grade is 1,003.8/5 = 235.3 g/t.

Although a metal equivalent accounts for the lead and zinc in the deposit, it depends on grades, recoveries, and prices, and therefore it depends on how these parameters vary during the mine life.

METAL PRICES

The process for determining the current price of a metal is really no different from the processes used to determine the price of anything: Connect as many buyers and sellers of the metal as possible and provide a mechanism by which sellers can ask a price and buyers can offer a price. Given enough time, participants will agree to a price. The more participants, the more information becomes available to inform asking and offering prices.

There are four methods by which metal prices are determined. They are listed here in order of increasing number of participants:

1. *Producer prices.* Producers set their price by taking account of costs, potential markets, and levels of competition. Producer pricing is common for industrial minerals where transportation costs are high. There is usually little room for negotiation of the price.
2. *Negotiated prices.* Prices are determined by direct negotiation between a buyer and a seller.
3. *Independent pricing.* Prices are determined by sources that are neither buyers nor sellers of metals. The prices are averages of prices of actual transactions

between producers, consumers, and metals traders. Examples include gold, platinum, magnesium, titanium, iridium, aluminum, and uranium.

4. *Commodity exchange pricing.* Prices are determined by transactions between dealers who are representatives of metal buyers, sellers, and metal traders.

Examples of negotiated pricing, independent pricing, and commodity exchange pricing are described in the following sections.

Negotiated Pricing

A buyer and seller of a metal can enter into a private contract that determines the conditions and price for sale of the metal. This is common in long-term contracts for iron ore, metal concentrates, or metal products. There can be terms in a contract that depend on the price set by a commodity exchange (described in the "Commodity Exchange Pricing" section). Two examples are a base metal contract (a smelter contract) and a precious metals contract (a refinery contract).

Smelter Contracts

Smelter contracts are agreements between a mine that produces a base metal concentrate and a smelter that processes the concentrate into pure metal. Figure 7.4 shows the components of a copper smelter contract. A smelter contract contains details concerning:

- How the mine will be paid for the principal metal in its concentrate
- How the mine will be credited for other desirable metals in the concentrate (e.g., by-products such as gold)

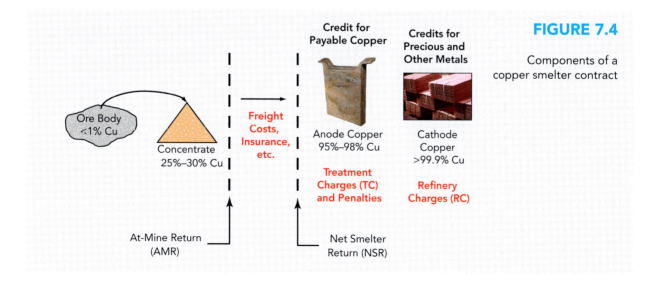

FIGURE 7.4

Components of a copper smelter contract

- What penalties will be applied for materials that affect the performance of the smelter (e.g., antimony, arsenic, bismuth, moisture)
- How the delivery is to be made
- The manner in which check assays of the concentrate will be done

The payment received once the treatment and refining charges (TC/RC) are applied is often called the *net smelter return* (NSR) per ton of concentrate, or NSR_c. The mine is responsible for transportation, insurance, and agents' costs. These costs, which are relatively small compared to TC/RC, are subtracted from the NSR to obtain the *at-mine return* (AMR).

The NSR or AMR can be between 60% and 90% of the value of the metal shipped to the smelter, depending on the demand for the metal and the demand for the concentrate. This is known as the *percentage paid*. The value of a metric ton of concentrate that leaves the mine can be reasonably approximated by applying a percentage paid to the value of the contained metal. Although the details of a smelter contract are confidential, an estimate of the value of the percentage paid can usually be obtained by inquiry.

Knowing NSR_c and the concentration factor CF (see Chapter 3), the NSR per ton of ore, NSR_o, can be computed as

$$NSR_o = \frac{NSR_c}{CF}$$

Refinery Contracts

Refinery contracts are much simpler, as shown in Figure 7.5; they are similar for other precious metals. An impure metal (doré in the case of gold) is produced at a mine and refined at an independent refinery. Several refineries are located in major centers around the world. The refinery contract prescribes how the refinery will assay, treat, and pay for the doré.

FIGURE 7.5

Gold metal processing and refining

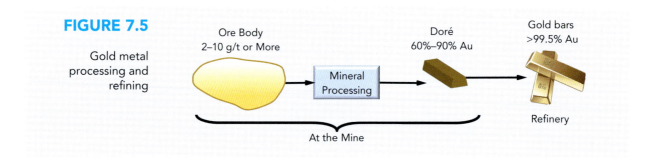

Typical terms of a gold refinery contract are:

- A treatment charge of US$0.80 to US$1.20/oz is levied, depending on current market conditions.
- The refinery typically pays the mine for 98% to 99.95% of the gold contained in the doré, depending on market conditions.
- Penalties are applied for deleterious elements such as iron, lead, tellurium, and nickel.
- The refinery will pay between 95% and 99% of the silver content of the doré.

The details of refinery contracts concern the procedures established for weighing and assaying. Security measures, delivery dates, disposition of refinery waste, and transportation of the doré are all dealt with in a refinery contract.

Independent Pricing

Independent pricing is done by parties who are essentially observers of the market. Examples are S&P Global's Platts Metals Daily (www.spglobal.com) and Fastmarkets (www.fastmarkets.com), who provide prices for precious metals, ferrous and nonferrous metals, and metal alloys and steel. The London Bullion Market Association (LBMA; www.lbma.org.uk) and the London Platinum and Palladium Market (LPPM; www.lppm.com) set prices for precious metals.

Buyers and sellers of uranium negotiate prices privately. Each month independent organizations, such as the UxC (www.uxc.com) and TradeTech (www.uranium.info), record average prices.

Iron ore prices used to be based on a long-term benchmark price established by a small group of producers and steelmakers. Recently, there has been a shift to more short-term pricing based on transactions within a larger group of companies and there are efforts to develop a futures market. See "How the Iron Ore Market Works" at Investopedia.com (2022) for a discussion of the iron ore market.

Germanium is a by-product of zinc ore processing. The market for germanium might be called "thin," that is, a relatively small number of buyers and sellers compared to other metals such as copper or iron. But the metal is considered essential in the semiconductor industry. An example of how it is traded through independent pricing is described in the box on the next page.

The LBMA sets prices for gold and silver using an electronic auction administered by ICE Benchmark Administration (www.ice.com/iba/). Direct participants in these auctions include banks, bullion dealers, and precious metal refiners. There are also indirect participants who are clients of direct participants. Auctions are held

GERMANIUM PRICES DROP ON SURPLUS MATERIAL IN CHINA

Germanium prices in Europe fell on Wednesday, June 14, 2018, as a result of surplus material available in China and low demand from end users.

Metal Bulletin assessed germanium metal at US$1,560–$1,660 per kg on an in-warehouse Rotterdam basis on Wednesday, down 5.3% from US$1,650–$1,750 per kg the previous week.

"A European trader offered us 50 kilograms of germanium at $1,575 per kg, but we didn't buy, so I guess they sold it at a lower price," a supplier said.

Source: Metal Bulletin 2018

at specific times of the day (gold at 10:30 and 15:00, silver at 12:00, all London time). Bids are entered by direct participants at 30-second intervals. Bidding continues until the amounts of metal to be bought and sold balance within thresholds (10,000 oz for gold; 500,000 oz for silver). If there is no balance, the price is changed and the process begins again until a balance is found. More details can be found at the ICE LBMA gold and silver price webpage (www.ice.com/iba/lbma-gold-silver-price).

The LPPM uses an electronic auction administered by the London Metal Exchange (LME) to set platinum and palladium prices daily. See the LME precious metals webpage (www.lme.com/en/Metals/Precious).

Gold is perceived as a safe store of value regardless of economic activity. For that reason, gold prices tend to increase when economic conditions are poor (e.g., inflation) or uncertain. Figure 7.6 shows time histories of the gold price and annual change in the Consumer Price Index (CPI), a measure of inflation. The graph shows that during the late 1970s, gold prices increased when oil prices escalated, leading to inflation. From about 1995 to 2002, strong economic growth made currency more attractive than gold and central banks were selling their gold, causing an increase in supply and a decrease in the price of the metal. From about 2003 to 2020, inflation decreased, but a combination of political and market uncertainty, as well as the economic effects of the pandemic, drove up the price of gold.

Commodity Exchange Pricing

Metals are traded at many exchanges but the two major exchanges are the LME (www.lme.com) and the Chicago Mercantile Exchange (now called CME Group; www.cmegroup.com/markets/metals.html). CME Group trades in futures and options contracts for metals, sometimes known as *derivatives*. The LME trades in

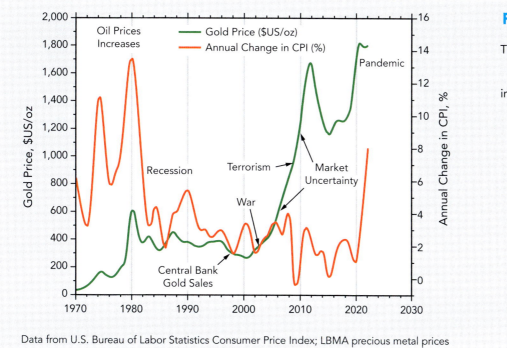

FIGURE 7.6

Time histories of gold price and annual change in the Consumer Price Index

Data from U.S. Bureau of Labor Statistics Consumer Price Index; LBMA precious metal prices

aluminum, cobalt, copper, molybdenum, lead, nickel, tin, zinc, and steel products as well as in futures and options for these metals.

Daily (or "spot") prices at both exchanges are determined by electronic trading. The LME also uses a continuous open-outcry auction carried out by traders physically located in a "ring" where one metal is traded for 5 minutes at particular times of the day. Telephone trading is also available at the LME.

Commodity exchanges have warehouses around the world where a physical supply of a metal is stored. A measure of the supply of a metal is the amount available for purchase in these warehouses on a particular day—the stock. Traders know this supply and also recognize any constraints on supply such as smelter or mine shut downs. Thus, they know quite a lot about the market and bid or ask a price on that basis. Prices should therefore reflect the available information, one indicator of an efficient market (Downey et al. 2023).

The relationship between copper stock and price at the LME for the period 1990–2022 is shown in Figure 7.7. With some exceptions, the expected relationship is apparent such as when stock increases, price decreases and vice versa. The exceptions and the price volatility (large variation in a short period) are interesting and may be related to the "financialization" of the copper market where copper is being

FIGURE 7.7

Copper price and copper stock

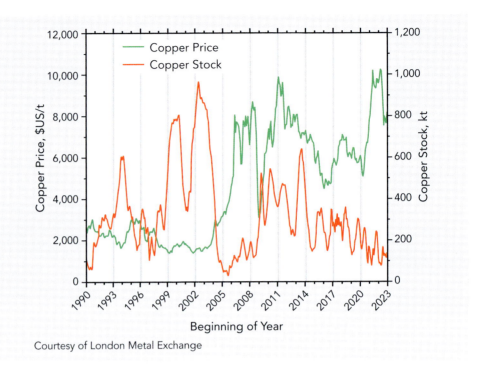

Courtesy of London Metal Exchange

used as an investment, that is, part of the asset portfolios of large investors. Trading in these portfolios in response to influences outside the copper market leads to price fluctuations that are not related to stock levels.

Further details about the efficiency of metal markets and their financialization can be found in Park and Lim (2018), Cheng and Xiong (2014), and Kesselring et al. (2019).

RESOURCES AND RESERVES

After the geometry of the ore deposit is defined by detailed exploration, mine design can proceed. An objective of the design is to classify portions of the deposit according to whether they can be *economically* mined and processed. This constitutes a legal definition of *ore* and is what is publicly reported to shareholders of the company that owns the rights to the deposit.

There are two basic classifications of known mineral deposits. One is *resources* and the other is *reserves*. Commercial extraction of resources is potentially feasible but the feasibility has not been confirmed. By contrast, it is feasible to extract reserves economically and legally. Another way of defining reserves is "one can go out today and extract them."

Reserves and resources are further classified as shown in Figure 7.8. This is known as a *JORC diagram* because it was devised by the Joint Ore Reserves Committee (JORC) of the Australasian Institute of Mining and Metallurgy (JORC 2012). There are two dimensions by which resources and reserves are subclassified. One is geological knowledge and the other is "modifying factors." The modifying factors are listed in Figure 7.8, but once these are considered and dealt with, resources become reserves. Consideration is actually not enough—there must be action, including design, permitting, and construction of a mine capable of producing.

All mines start with exploration results, which could be some combination of geological, geophysical, and geochemical surveys. After those methods are employed, it is mainly drill-core samples that provide more geological knowledge. As confidence in that knowledge increases, resources go from inferred, to indicated, to measured; and reserves go from probable to proven. These designations can be applied to parts or volumes of the same mineral deposit.

Generally, after modifying factors have been considered, indicated resources become probable reserves and measured resources become proven reserves. Inferred resources cannot be used to estimate reserves. If there is a low level of confidence in the ability to mine and process the measured resources or some other uncertainty associated with the modifying factors, it is possible for some of the measured resources to become probable reserves, as denoted by the dashed red line in Figure 7.8. The vertical axis indicates an increasing amount of geological knowledge. However, note that the dashed line does not represent a change in the level of geological knowledge.

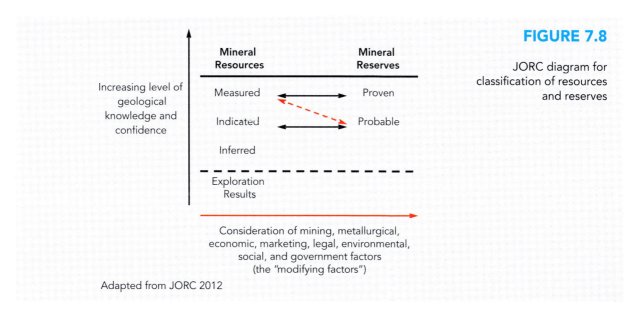

FIGURE 7.8

JORC diagram for classification of resources and reserves

No general rules or principles govern the transition from one type of resource or reserve to another. The designation is made by a geologist based on knowledge of the type of ore body, geological structure, and professional experience. An example is discussed next.

The Minto Deposit—An Example of Resource Classification

The Minto deposit is located in Yukon, Canada. Initially the reserves were small, about 9 Mt (million metric tons), but the grades were high: 1.78% copper, 0.62 g/t gold, 7.3 g/t silver, all contained in chalcopyrite and bornite mineralization. The mine began production in 2007 and its expected life was 7 years. Since then, more reserves were found and the mine life was extended to 2022.

The original exploration work done for the Minto deposit provides an excellent illustration of how geological and drill-hole data are used to classify resources. Several drilling programs had been conducted in the area of the deposit prior to 1993. High-grade (~2%) copper mineralization was found in the drill cores. The drilling results also showed that the mineralization was in sub-horizontal zones at depths between 100 and 200 m that could be traced in a roughly north–south direction for more than 1 km.

Figure 7.9 shows a plan view of all drill holes from the period before 1993. Close spacing of the holes can be seen, some less than 20 m apart. Continuity of the mineralization was assumed between most of the drill holes. From this figure, it can be seen that the spatial density of drill holes determines whether a resource is classified as measured, indicated, or inferred. A high spatial density of drill holes is needed to classify a measured resource, whereas an indicated resource zone contains at least one drill hole. An inferred resource zone contains no drill holes, but it is an extension of an indicated ore zone in the north–south direction.

Although the orientation of the extension of an inferred resource zone is suggested by the known north–south orientation of the mineralized zone, the distance of the extension is dictated by drill-hole density. For example, the inferred and indicated resources circled with the dashed red line in Figure 7.9 are oriented north–south, even though there is only one drill hole at the center of the indicated resource, but no continuity is assumed between these resources and the other resources to the east.

Cutoff Grade

Reserves must be extracted economically. For pure metals (e.g., gold), this means the following inequality must be satisfied:

$$\frac{price}{unit\ mass} \times grade \times recovery > \frac{total\ costs}{unit\ mass}$$

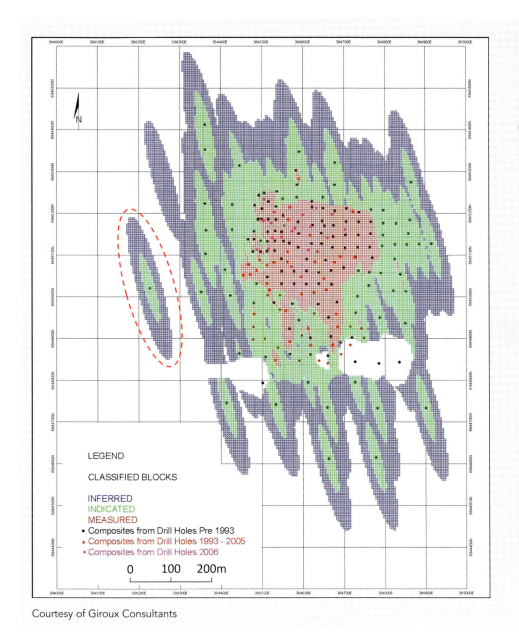

FIGURE 7.9

Plan view of the drill holes (shown as dots) used to make the mineral resource estimate for the Minto mine, Yukon, Canada

LEGEND

CLASSIFIED BLOCKS

INFERRED
INDICATED
MEASURED
• Composites from Drill Holes Pre 1993
• Composites from Drill Holes 1993 - 2005
• Composites from Drill Holes 2006

0   100   200m

Courtesy of Giroux Consultants

where total costs are the expected costs of mining and/or processing. At equality:

$$\text{cutoff grade} = \frac{\text{total costs}}{\text{price} \times \text{recovery}}$$

The *cutoff grade* is the lowest grade that can be mined and processed economically. Recovery is usually constant, but costs and price change over time so that the cutoff grade will change during the life of a mine. For base metals, a *cutoff* is defined in

FIGURE 7.10

Block model of ore body showing economic and uneconomic ore blocks

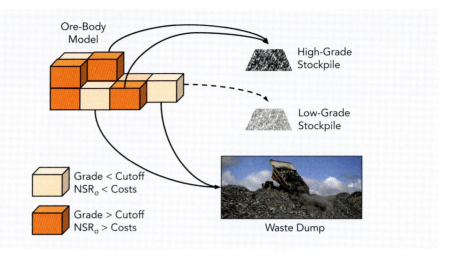

terms of the net smelter return per ton of ore, NSR_o. If NSR_o is less than total costs, the rock may be considered waste.

A rock mass with grade less than the cutoff grade, or that costs more to mine and process than the potential revenue, is usually considered to be waste, but it might be stockpiled in a low-grade stockpile to be processed later. The decision is depicted in Figure 7.10, which shows an ore-body model consisting of blocks (each perhaps 10–20 m on a side) of known average metal grade. The expected costs, grade, and metal recovery of mining and processing each block can be used to determine whether or not the block is economic in terms of a cutoff grade or NSR_o. This information is used to establish where the rock in each block goes, to a high-grade stockpile (typically next to the concentrator), to a low-grade stockpile, or to the waste dump.

Different assumptions for cutoff grade result in different amounts of available economic resources or reserves within a deposit. Figure 7.11 shows an example of a graph of the percentage of reserves available versus grade for a copper deposit that has an average grade of about 0.5%. If the cutoff grade is low, most of the deposit becomes economic. However, the amount of economic material quickly decreases as the cutoff grade approaches the average grade. This is because the distribution of grades in the ore deposit is biased toward lower grades. This is a typical property of metal grade distributions. (See Appendix B.)

Given an assumed price and the sum of the expected mining and processing costs, a cutoff grade can be established for a block of ore in the ground to be mined (the mining cutoff). Similarly, using only the expected processing costs, a cutoff grade

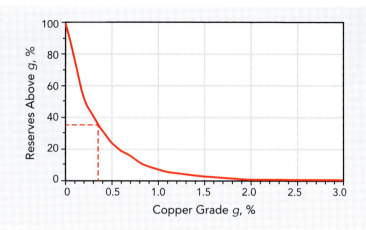

FIGURE 7.11

Percentage of reserves available versus grade, g

can be established for a stockpile to be processed (the mill cutoff). However, any calculation of cutoff grade must account for the effects of errors in grade estimation, dilution, mining losses, process plant losses, and royalty agreements. This has implications for estimation of reserves, mine planning, and mine life.

The multiplying factor in cutoff grade, f = costs/recovery, results in two cutoff grade scenarios:

1. If the estimated cutoff grade is too low, more of the ore deposit becomes economic and the mine life is longer. However, the factor f is low, meaning that either the estimated costs are too low or the estimated recovery is too high and the amount of metal in the deposit that can be recovered economically is overestimated. The production capacity of the resulting mine and processing plant would be too large and the expected production will not be achieved, leading to financial losses. To maintain the operation in this case, the only strategy is to raise the cutoff grade and shorten the mine life.

2. Conversely, if the estimated cutoff grade is too high, less of the ore deposit is considered economic and the mine life is shorter. However, the factor f is high, meaning that either the estimated costs are too high or the estimated recovery is too low and the amount of metal that can be recovered economically is underestimated. The production capacity of the resulting mine and processing plant would be too small. If the cutoff grade is lowered and more economic ore becomes available, the operation would not have the capacity to mine and process it. An advantage of this situation results if there is the possibility to invest and add more capacity to the operation, but this depends on the amount of flexibility built in the original design to accommodate such an expansion.

REPORTING OF RESOURCES AND RESERVES

The amount of resources and reserves is obviously important to the financial health of a mining company (no reserves, no mining). For this reason, mining companies are always looking to acquire more reserves, either by exploration for resources that could be developed into reserves or by acquisition of a company with resources or existing reserves. This raises the issue of how resources and reserves are reported and who does the reporting. Investors in mining companies also have an interest in how resources and reserves are reported.

JORC Code

The Australasian Institute of Mining and Metallurgy formed JORC in 1971 to develop definitions and reporting regulations. It had become evident that there was no common framework and language available to describe potential ore deposits. In 1989, JORC released the first edition of the JORC Code, which contained the original JORC diagram shown in Figure 7.8 and which defined the terms resources and reserves; it further clarified the terms inferred, indicated, and measured resources, and probable and proven reserves. The most recent version of the code was issued in 2012 (JORC 2012).

The JORC Code specifies standards for public reporting of resources and reserves as well as other technical information associated with mineral deposits. Such reporting could take many forms, including annual or quarterly company reports issued, press releases, website or social media postings, and public presentations. Care must be taken with oral comments made in public situations such as elevators or office social events.

An important requirement of the JORC Code is that a competent person (CP) must be designated and identified as the originator of the analysis that led to statements in public reports. The code does not specify procedures to be used by the CP and, given the variety of issues and possible methods, it cannot do so. The decision as to what procedures to use is left up to the professional judgment of the CP. Typically a CP is an individual who is a registered engineer or geoscientist with at least 5 years of experience relevant to the subject matter of the disclosure.

Initially, JORC was concerned with the reporting of resources and reserves, but over time, the issue of feasibility of mining projects became increasingly important, mainly because of the large costs and risks associated with these projects. Accordingly, a framework and standards for technical reports on feasibility was developed. Three types of technical report are identified in JORC 2012:

1. *Scoping Study.* This includes an economic analysis of the *potential* viability of mineral resources using order of magnitude estimates of costs and production rates and, where appropriate, the marketability of the mineral product. A scoping study may include inferred resources in the economic analysis but must clearly state how inferred resources are used in all analyses. The purpose of a scoping study is to demonstrate that further investigation into project feasibility is justified.

2. *Prefeasibility Study* or *Preliminary Feasibility Study* (PFS). A range of methods for mining, mineral processing, and waste management is studied to establish a preferred set of options. A financial analysis is performed using reasonable assumptions about the modifying factors (e.g., assuming that permits can be obtained). A PFS should provide sufficient information to determine if some portion of the defined resources can be converted to reserves.

3. *Feasibility Study* (FS), in which all geological data are included, modifying factors are assessed in sufficient detail, and a detailed financial analysis is performed *to allow a decision on financing the project. Providing owners and investors with sufficient information to enable a decision on financing* is the purpose of an FS. Although there may be design information in an FS, a contractor tasked with building all or part of a mine would use the study as a starting point for detailed design.

FSs are comprehensive documents. Depending on the size of the project, the cost of a full FS can be a few million dollars. Several CPs may be involved in an FS, one for each of the relevant modifying factors: geology, mining, processing, waste management, environment, economics, and so forth. A lot of useful information is contained in FSs concerning how mines are planned and designed. Copies of FSs or technical reports from Canadian public companies can be downloaded at www.sedar.com and those from U.S. companies can be downloaded at sec.gov/edgar.

Reporting Standards

In 1994, the Committee for Mineral Reserves International Reporting Standards (CRIRSCO) was formed as part of an international effort to standardize reporting of mineral resources and reserves (crirsco.com). Its members consist of national mineral resource reporting organizations (NROs) in several countries on all continents (CRIRSCO, n.d.-b). The need for standardization came about as a result of increased globalization of the mining industry, where a mining company based in one country would seek to develop mining projects in one or more other countries.

Each of these NROs adopted the JORC Code for reporting resources and reserves and for technical reports. In 2019, CRIRSCO issued a standard template for technical reports (CRIRSCO, n.d.-a) that was based on the technical report standards used by some of the NROs. As a result, the format of technical reports issued in different countries is similar.

All public mining companies whose shares are traded on a stock exchange in a country are required to publicly disclose scientific and technical information about their mineral projects according to a set of rules and standards developed by exchange regulators of that country. The rules and standards describe such things as the conditions under which a technical report must be filed, what kind of resources can be included in a technical report, and who can be called a CP. Table 7.1 provides a list of exchanges and links to their standards of disclosure. Unless otherwise specified in the standards, it is assumed that technical reports will follow the CRIRSCO template.

TABLE 7.1

Securities regulators and standards of disclosure for mining companies

Regulator	Reporting Standard	Web Address
Australian Securities and Investments Commission	Chapter 5: Additional Reporting on Mining and Oil and Gas Production and Exploration Activities in Australian Securities and Exchange (ASX) Listing Rules	www.asx.com.au
Canadian Securities Administrators	NI 43-101 (Standards of Disclosure for Mineral Projects)[*]	www.osc.ca
Johannesburg Stock Exchange (JSE)	Section 12, JSE Limited Listings Requirements	www.jse.co.za
U.S. Securities and Exchange Commission	17 CFR Part 229 Subpart 229.1300— Disclosure by Registrants Engaged in Mining Operations[†]	www.ecfr.gov

[*] Scoping Study = Preliminary Economic Assessment; Competent Person = Qualified Person.

[†] Scoping Study = Initial Assessment.

ECONOMIC ANALYSIS

One important component of technical reports is an economic analysis of the proposed project. The components of an economic analysis are shown in Figure 7.12a. This requires making estimates of the costs to build the project (capital costs), costs of operation (mining, processing, waste management), and revenue over the life of the mine. Sustaining capital costs (e.g., equipment replacement) may be incurred during the life of the mine. Sometimes taxes and royalty payments to governments are included in the analysis. Financing arrangements might be used to provide the

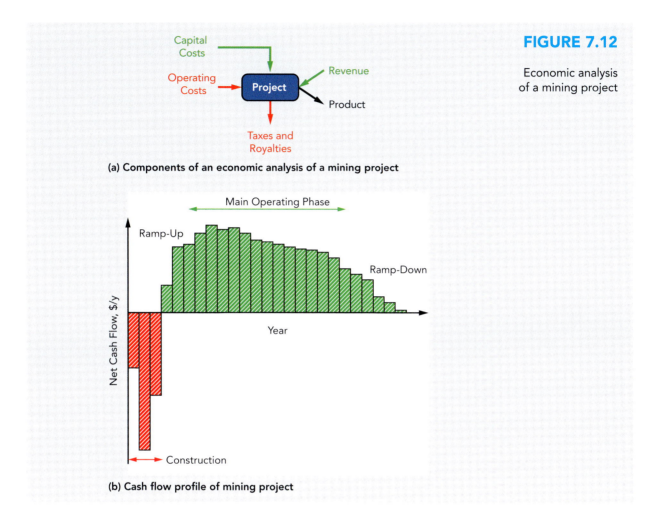

FIGURE 7.12

Economic analysis
of a mining project

(a) Components of an economic analysis of a mining project

(b) Cash flow profile of mining project

capital costs, but it is usually assumed in the analysis that capital costs are provided by investors and cash (all equity financing).

The net cash flow for a particular period is the difference between revenue and the sum of capital costs, operating costs, taxes, and royalties. Typically, as shown in Figure 7.12b, the net cash flow divides a project into a construction phase (zero revenue), a ramp-up period as production reaches capacity, an operation phase where revenues fluctuate and sustaining capital costs are incurred, and a ramp-down period where reserves decrease.

Naturally, there are significant uncertainties associated with these revenue and cost estimates. One objective of the analysis in a technical report is to quantify the range of these uncertainties.

WHERE DOES THE MONEY COME FROM?

The amount of capital required to build a mining project can range from several hundred million to several billion dollars. One source of capital is available cash, but often this must be supplemented by issuing debt securities (selling promises to pay) or with proceeds from the sale of shares (selling equity in the company). There are advantages and disadvantages to each of these sources. Issuing debt sends a signal that the company is financially able to meet debt obligations. Issuing shares for sale could dilute the value of existing shares (higher supply, lower price). Often a balance must be struck.

Metal streaming is an interesting financing scheme in which a mining company agrees to sell a percentage of its metal output at a reduced price over all or part of the life of a proposed mine to an investment company sometimes known as a "metal streamer." In return, the metal streamer provides part of the capital required to build the project. Metal streaming has advantages and risks for both parties to the streaming agreement, but both will want maximum production from the mine to be achieved.

Further Suggested Reading

If you want to learn more about the world's metal supply and trading system, here are some recommended books on the subject:

- *An Introduction to Trade and Commodity Finance: With Examples from the Trade in Metals* (De Jong 2021)
- *The Rare Metals War: The Dark Side of Clean Energy and Digital Technologies* (Pitron and Jacobsohn 2020)
- *The World for Sale: Money, Power, and the Traders Who Barter the Earth's Resources* (Blas and Farchy 2022)
- *Volt Rush: The Winners and Losers in the Race to Go Green* (Sanderson 2023)

REFERENCES

Blas, J., and Farchy, J. 2022. *The World for Sale: Money, Power, and the Traders Who Barter the Earth's Resources.* New York: Oxford University Press.

Cheng, I-H., and Xiong, W. 2014. Financialization of commodity markets. *Annual Review of Financial Economics* 6:419–441.

CRIRSCO (Committee for Mineral Reserves International Reporting Standards). n.d.-a. *CRIRSCO International Reporting Template for Exploration Results, Mineral Resources and Mineral Reserves.* www.crirsco.com/template. Accessed November 2023.

CRIRSCO (Committee for Mineral Reserves International Reporting Standards). n.d.-b. CRIRSCO members. www.crirsco.com/members/. Accessed November 2023.

De Jong, G. 2021. *An Introduction to Trade and Commodity Finance: With Examples from the Trade in Metals.* Utrecht, The Netherlands: Eburon Academic.

Downey, L., Scott, G., and Velasquez, V. 2023. Efficient market hypothesis (EMH): Definition and critique. www.investopedia.com/terms/e/efficientmarkethypothesis.asp. Accessed November 2023.

Investopedia. 2022. How the iron ore market works. January 24. www.investopedia .com/articles/investing/030215/how-iron-ore-market-works-supply-market-share.asp. Accessed November 2023.

JORC (Joint Ore Reserves Committee). 2012. *The Australasian Code for Reporting of Exploration Results, Mineral Resources and Ore Reserves (The JORC Code)*. Gosford, New South Wales: JORC. www.jorc.org. Accessed December 2023.

Kesselring, R., Leins, S., and Schulz, Y. 2019. Valueworks: Effects of financialization along the copper value chain (Working Paper). Geneva: Swiss Network for International Studies.

Metal Bulletin. 2018. Germanium prices drop on surplus material in China. June 14. www .fastmarkets.com/insights/germanium-prices-drop-on-surplus-material-in-china/. Accessed November 2023.

NI 43-101. 2002. Canadian National Instrument. *Standards of Disclosure for Mineral Projects*. Expanded in 2011. https://www.osc.ca/en/securities-law/instruments -rules-policies/4/43-101/ni-43-101-standards-disclosure-mineral-projects-form -43-101f1-technical-report-and-related. Accessed November 2023.

Park, J., and Lim, B. 2018. Testing efficiency of the London Metal Exchange: New evidence. *International Journal of Financial Studies* 6(1):32.

Pitron, G., and Jacobsohn, B. 2020. *The Rare Metals War: The Dark Side of Clean Energy and Digital Technologies*. New York: Scribe US.

Sanderson, H. 2023. *Volt Rush: The Winners and Losers in the Race to Go Green*. London: Oneworld.

United Nations. 2023. Population. www.un.org/en/global-issues/population. Accessed November 2023.

World Bank. 2023. World Bank open data. https://data.worldbank.org/. Accessed November 2023.

A Future of Mining

It is relatively easy to predict what will happen in the mining industry in the next 5 to 10 years. But what will a mine look like in 20 years? 50 years? 100 years? This is much harder to predict, but there are some indications for where developments and changes will occur and some rather strong indications of what has to happen for mining to continue.

This chapter discusses these indicators and their significance and describes one rather extreme scenario. However, the idea is not to promote this scenario, but to start a dialogue and to encourage thinking about what the future of the industry might look like.

Notice that the title of this chapter is "**A** Future of Mining," not "*The* Future of Mining." There could be many possibilities and it is fun to think about these—the only limitation is your imagination.

DRIVERS OF INNOVATION IN MINING

The mining industry is perceived as a slow innovator. However, Bartos (2007) used productivity data to show that the rate of innovation in metal mining companies is comparable to rates associated with manufacturing and other mature industries, but not as fast as the innovation rates of advanced technology industries.

Although innovation occurs in the industry and in associated organizations, the rate of adoption of these innovations is not large. This is likely the source of the perception that mining is a slow innovator and too conservative. However, there is good reason for this—risk. The enormous financial, environmental, and social consequences of some innovative device or process not working as expected, or failing to work at all, are significant deterrents to its adoption.

In the future, the rate and adoption of innovations in mining will have to be faster. The reason is that there are many drivers of innovation in the industry, but they are actually in the form of constraints on the current mining paradigm. Overcoming these constraints will be challenging, requiring many innovations, the use of a wide array of different technologies, and "out-of-the-box" thinking. However, it can be done.

Three of these drivers are discussed in the following sections.

Sustainable Mining or Sustainable Supply of Metals and Materials?

Sustainable mining might sound like an impossibility and, in the sense that it involves depletion of a finite resource, mining, as we have come to know it, is not sustainable. In Chapter 6, sustainability is defined as meeting the current needs and demands of society without compromising the needs and demands of future generations. However, if the industry cannot supply the metals and materials needed by society, then the ability of future generations to meet their needs will most certainly be compromised.

In different ways, geology, current mining and processing technologies, the organization of the metal supply system, economics, geopolitics, and the environmental and social footprints of mining each constrain the ability of the industry to provide metals and materials. Achieving a sustainable supply of metals and materials is a strong driver for developing changes in technology and in the metal supply system to circumvent these constraints.

Mineral Deposit Discovery Rates

A web search with the keywords "mineral deposit discovery rates" will result in about 18 million websites, many of which document a decline in the rate of mineral deposit discovery since the early part of the 21st century. Some sites also mention a decline in exploration expenditures by major mining companies. Smaller ("junior") mining companies do exploration work, but the financial support for these companies is mainly venture capital, which depends on the risk tolerance of the mining investment community and varies over time.

Another reason for the low discovery rate could be the assumption that models of ore formation currently used to find deposits will continue to be representative. There are other perspectives on models of mineral deposit formation. For example, in a report about a nationwide geological exploration initiative, the Australian Academy of Science (2012) states that ore formation systems are characterized by processes that occur on a large spatial scale. The report suggests that the focus of current exploration is too limited in scale and that it is necessary "to understand the larger mineral system and its *measurable characteristics* on various scales so as to predict and detect deposits within."

A similar perspective is provided by Richards (2013) who reviewed giant porphyry copper-molybdenum-gold and epithermal gold-silver deposits and found that the large accumulation of metal in such deposits is the result of a fortuitous coincidence of common geological, physical, and chemical processes. Exploration for such deposits must therefore seek the *distinct conditions* or indications of enhanced metal accumulation.

There are further constraints on developing a mine given a delineated mineral resource. The availability of funds, political issues, and market conditions combine to make the likelihood that the mineral resource becomes a mine about 50% and to cause the time to develop the discovery into a mine to be at least 10 years and likely more (Schodde 2014).

A result of the difficulty in finding and developing major ore deposits is that lower-grade and possibly smaller deposits will become potential targets. If this is the case, then alternative mining and mineral processing methods will have to be developed. The mineralogy of small or low-grade deposits can be complex so that the ore must be ground to very fine sizes to liberate the desired minerals. This leads to higher energy requirements and higher costs, prompting a need for improved methods of grinding.

Or should we ask whether there are alternatives to mining and grinding ore?

Metal Demand

Copper is widely used in the global economy and its demand is a leading indicator of economic growth (ICSG, n.d.). The ability to supply copper is therefore of considerable interest. Figure 8.1 shows a scatterplot of copper grades versus reserves or resources for several known deposits, depending on whether the deposit is an operating mine or not. This illustrates a challenge for copper supply. Most of the

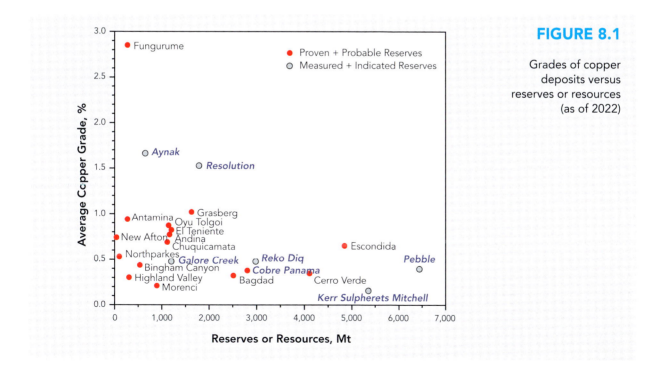

FIGURE 8.1

Grades of copper deposits versus reserves or resources (as of 2022)

operating mines shown in Figure 8.1 have short remaining lives, 10–20 years. To replace these reserves, the resources shown will have to be developed into reserves, that is, made into operating mines. However, these resources may take years to develop and pose significant technical challenges. For example, the Resolution deposit east of Phoenix, Arizona (United States), lies at a depth of 2 km where the temperature is 80°C. The other resources are in remote areas with difficult physical conditions. Innovation will be required to overcome the challenges.

Generally, as income and population grow, so does the consumption of products and systems that contain metals (Baffes and Nagle 2022). For example, demand for machines will increase as people with more income perceive an opportunity to do less manual labor. Urbanization, or the growth of population concentration in cities, causes increased demand for transportation systems and infrastructure, both of which require metals. But the increase in metal consumption with income is not linear and can vary between countries. Also, at high-income levels, the market for products containing metals becomes saturated (formally known as the "people have enough stuff" condition) so that the demand for metals may exhibit little change, or even decline, with income (Baffes and Nagle 2022).

Assumptions about urban and industrial growth, and the corresponding growth of demand for products, such as vehicles, appliances, and equipment, form the basis of a future economic growth scenario. Knowing the metal intensity of physical entities in the scenario (e.g., kilograms of copper per car), the future demand for metals in that scenario can be estimated.

The goal of the Paris agreement (UNFCCC 2016) is to take measures to maintain the global atmospheric temperature increase at 1.5°C above preindustrial (1850–1900) levels. To achieve this, by 2050, all human-derived carbon dioxide emissions will need to be matched by carbon dioxide removal using technologies such as direct air capture and storage (IPCC 2018). This is called the *net zero emissions* (NZE) scenario.

In the NZE scenario, clean technologies, such as solar and wind power, will have to replace fossil fuel energy generation and supply 70% of global electricity demand by 2050 (IEA 2021). This will lead to additional demands for metals, independent of other demands. For example, solar power and wind turbine systems contain 2–6 t (metric tons) of copper per megawatt, and electricity grids with copper wiring will connect these power sources. Figure 8.2 shows the resulting estimated copper demand for clean technologies during the period 2022–2050.

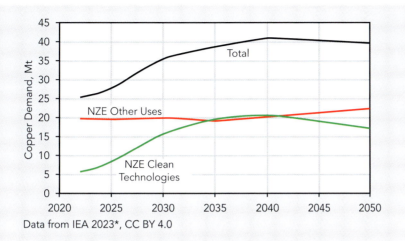

FIGURE 8.2

The estimated demand for copper during the period 2022–2050 (NZE = net zero emissions)

*The graph is not endorsed by the International Energy Agency (IEA) in any manner; the author is solely liable and responsible for it.

Vehicles, appliances, and industrial equipment, collectively known as "other uses," will be present in the NZE, and the estimated copper demand for these other uses during the period 2022–2050 is also shown in Figure 8.2. The total copper demand in Figure 8.2 is the sum of that for other uses and that for clean technologies.

The demand for copper used in clean technologies decreases after 2040 as clean energy generation systems are installed and meet the demand for electricity. However, using the current mining and metal supply paradigms, it will be impossible to supply the total demand past the year 2030 unless some difficult decisions and trade-offs are made. Demands for other metals, such as nickel used in electric car batteries, will also be difficult to supply.

Society needs innovation to find additional and alternative sources of these metals and to efficiently extract the metals from them.

PATHS FOR INNOVATION

The return on equity for a company is given by the product of two factors: *operating productivity* and *capital productivity*. The following proportionality can be written:

$$\text{return on equity} \propto \underbrace{\frac{\text{revenues} - \text{costs} - \text{taxes}}{\text{output}}}_{\text{operating productivity}} \times \underbrace{\frac{\text{output}}{\text{assets}}}_{\substack{\text{capital} \\ \text{productivity}}}$$

Output, which might be measured in pounds, metric tons, or ounces depending on the product, has been assumed to be equivalent to sales because all the output of a mine will be sold.

FIGURE 8.3

Autonomous (no driver) haul trucks at Rio Tinto's West Angelas mine in Australia

Photo © Rio Tinto 2015

Operating productivity and capital productivity provide the basis for three different paths for innovation in the mining industry. These are described in the following sections.

Path 1. Cost Reduction

One goal of a company is to maximize the numerators in the productivity factors. Mining companies have no control over revenue because prices are determined by external markets; they are *price-takers*. Tax regimes often favor the development and operation of mines, and taxes can be reduced by strategic management within a particular taxation system. Most mining companies maximize operating productivity by focusing on reducing operating costs.

One cost-cutting method is automation. The mining industry has been instrumental in the development of automated machines for drilling and for transport. Cost reduction is the main motivation. Transport of large amounts of rock is a tedious and repetitive operation, as well as costly in terms of labor and fuel. The other motivation is safety—it is inherently safer to use a process that physically removes humans from its operation.

In principle, it is possible to automate any part of the mining cycle. The part of the cycle that has received the most attention is ore transport because autonomous transport reduces labor costs and increases productivity. Autonomous haul trucks were originally employed at iron ore mines in the Pilbara region of Western Australia in 2008 (Figure 8.3) and are either in use or being considered at other open pit operations worldwide. Automated drilling is becoming more common in both open pit and underground mines. An online search with the keywords "autonomous mining machines" or "autonomous drilling machines" will result in many videos showing these machines in action.

However, costs of inputs to mining operations—notably equipment, reagents, and fuel—are increasing, and the industry has little control over these outlays. It is possible to organize mining operations to increase efficiency and minimize costs, but this often quickly reaches a point of diminishing returns. Consequently, to an increasing degree, mining companies have also become *cost-takers*; nevertheless, more innovations to avoid or reduce costs can be expected.

Battery-powered mining equipment such as load-haul-dump (LHD) units and haul trucks are currently in use or being tested for use (Volvo 2023; Teck 2021). The use of electricity eliminates the cost of diesel fuel. Electric machines operating underground require less ventilation because there are no emissions and little heat generation. In open pit mines, large fleets of small (e.g., 40 t), possibly autonomous, electric haul trucks would be used, and these do not require wide haul roads. Capital, operating, and maintenance costs of electric machines are considerably less than larger diesel machines.

Path 2. Capital Productivity

Using existing technology, it is possible to find ways of using assets more efficiently, that is, to get more out of assets and increase capital productivity. One interesting technology is ore sorting, where ore and waste minerals are separated underground or in the pit after blasting using differences in their physical properties (Robben and Wotruba 2019). Ore minerals typically have higher electrical conductivity or higher density and will reflect a particular kind of radiation differently. Thus, a conductive ore mineral will pass an induced electrical current, or a dense mineral will reflect more X-rays than it transmits. Combinations of sensors and mechanical systems are used to automatically separate the minerals. If this is done, more ore minerals can be transported by a mine shaft or a truck fleet, which results in more output from the same assets (i.e., greater capital productivity). The process has been implemented at some mines, resulting in 20%–50% waste rejection.

Path 3. What About the Assets in the Denominator?

Why not go further and develop radically different technologies that can do mining and mineral processing with large numbers of small, cheap, and even disposable assets? If this could be done, the number of assets and their corresponding value would be reduced, resulting in a larger return on equity. Dunbar and Klein (2002) proposed a variant of this idea. More recently, haul truck suppliers and some mining companies have started to test fleets of autonomous smaller trucks that can be electrified. (Search "fleets of small mining trucks" online.) A more extreme idea for small machines is described in the "Advanced Machines and Systems" section.

Although some serious out-of-the-box thinking would be needed to realize a situation in which small machines do mining, even farther out of the box is the idea of not using machines at all!

BIOTECHNOLOGY TO THE RESCUE?

The idea of not using machines to extract metals or other minerals seems farfetched. However, consider micrometer-sized microbes, which have been interacting with metals for about 2.5 billion years ever since the Great Oxygenation Event (Hazen and Ferry 2010) when cyanobacteria (blue-green algae) began to produce oxygen through photosynthesis. The increase in atmospheric oxygen created an environment in which microbes could thrive. Microbes either formed or decomposed minerals to produce carbon (food) or energy and these biochemical processes continued to the present. The result was an increase in the number of minerals from about 1,500 to the current number of 4,400.

The microbes would have had to adapt their genetic makeup to survive the many changes in their environment that occurred over 2.5 billion years. Microbial communities that evolve in and near mineral sources are therefore a rich source of genetic information. Collectively, these communities are called a *mineral microbiome*. The genetic information and pathways in this microbiome could be used to create synthetic or modified microbiomes that concentrate or sequester metals. The Mining Microbiome Analytics Platform (MMAP 2023) is a research project funded by the Canadian government, mining, and biotechnology companies, whose goal is to obtain this information by genetically sequencing samples from water bodies, soils, and mineralized surfaces at or near mine sites.

Low-cost, high-throughput, portable gene-sequencing tools, open-source device designs, collaboration, and crowdsourcing have made such a project feasible. A kind of "bio-prospector" community has developed that contributes to the knowledge base.

The mineral microbiome is a subset of Earth's microbiome (Earth Microbiome Project 2023; Gilbert et al. 2014; Thompson et al. 2017), but equally vast spatially. By way of comparison, it took about 13 years (1990–2003) to complete all the gene sequencing for the Human Genome Project (NHGRI, n.d.). However, the boundaries of that genome are known. The boundaries of the mineral microbiome are not as well-defined.

There are many references on applications of biotechnology to exploration, metal extraction, and to remediation and recovery from waste (Brune and Bayer 2012; Johnson 2015; Dunbar 2017; Levett et al. 2021; Kaksonen et al. 2020). The following examples illustrate the wide range of interactions between metals and microbes and the possibilities for their use.

Microbes Make Nuggets

Figure 8.4 shows some gold nuggets found in the Arizona desert using a metal detector. How did they get there? Are the nuggets the result of an accumulation of loose fragments worn away from the "mother lode"—a detrital process? This is not likely because gold is relatively heavy and not that mobile. Also, where's the mother lode?

FIGURE 8.4

Gold nuggets found in the Arizona desert using a metal detector

Courtesy of Bill Southern, www.nuggetshooter.com

Did the gold nuggets crystallize out of solution? That is also unlikely because to get into solution, it would have to oxidize to become an ion, and gold does not do that easily—it is a noble metal. Also, where's the water?

There has to be another way.

Gold nuggets are also found in the Australian outback, and their presence there raises the same questions about their origin. Frank Reith of the Commonwealth Scientific and Industrial Research Organisation in Australia decided to have a look at one such gold nugget under an electron microscope. He saw what looked like biofilms on the nugget, something a microorganism would produce. He applied a fluorescent blue dye that binds to the DNA molecules on the biofilm. An electron micrograph of the result is shown in Figure 8.5, which shows a 200 × 100 μm (micrometers) biofilm. A genetic analysis of the biofilm showed evidence of bacteria, a particular bacterium known by the Latin name *Ralstonia metallidurans*. This is interesting because metals are toxic to most bacteria (and to most living things if the concentrations are high enough). How did *R. metallidurans* survive in a toxic environment?

FIGURE 8.5

Electron micrograph of gold nugget showing a 200 × 100 μm biofilm stained by a fluorescent blue dye that binds to DNA

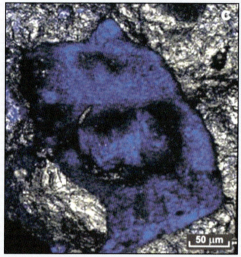

Source: Reith et al. 2006, © The American Association for the Advancement of Science

Experiments carried out by Reith et al. (2006) showed how *R. metallidurans* survived and made its growth possible. A gold chloride ion, $AuCl_4^-$, was added to a bacterial growth medium containing *R. metallidurans*. Initially, the growth of the bacterium was slowed because of the toxicity of the gold in solution, but after 72 hours, the growth rate increased. About 1.1 mg (milligrams) of gold had precipitated; that is, the solution of gold chloride was missing 1.1 mg of gold.

An electron micrograph of one cell of *R. metallidurans* from the growth medium revealed that the bacterium accumulated or sequestered the gold as a nanoparticle in its cell wall (Figure 8.6). Keeping the metal in its cell wall protected the bacterium from the toxicity and allowed it to grow. This work provided strong evidence that bacterial processes could contribute to the formation of secondary gold grains (i.e., nuggets) by sequestering the metal to detoxify their environment. Nature follows the adage: "Keep your friends close and your enemies closer."

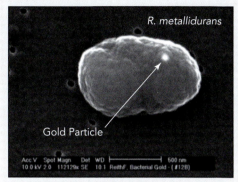

FIGURE 8.6

Electron micrograph of a cell of *R. metallidurans* grown in a medium in which a solution of gold chloride was introduced. The bacterium caused precipitation of a gold nanoparticle in its cell wall.

Source: Reith et al. 2006, © The American Association for the Advancement of Science

Termites and Metals

Some insects sequester gold and other metals. Throughout the central part of Australia, gold deposits are covered by a thick layer of transported soil. Few to no geochemical anomalies are observed at the surface in these soils, making detection of the gold deposits difficult and requiring expensive drilling programs to find zones of high mineralization. However, higher concentrations of metals may occur within the soil over a mineralized zone as a result of leaching processes below the water table or capillary action above the water table. As termites gather material for their nest, they burrow to depths of 1–4 m, and if they burrow through a soil containing high metal concentrations, small quantities of gold and other metals will accumulate in their bodies. As a result, anomalous concentrations of these metals will occur in termite mounds and in soils adjacent to the mounds, as shown by Stewart et al. (2012). Thus, the presence or absence of metals in a termite nest provide a means of better targeting drilling operations.

Methanogens

A methanogen is a microorganism that produces methane as a by-product of its metabolism. They are anaerobes; that is, they do not require oxygen. Instead, some methanogens use carbon dioxide as a source of carbon. Hydrogen, a waste product of the metabolism of other microorganisms, reacts with the carbon dioxide to produce methane:

carbon dioxide	+	hydrogen	$\rightarrow$	methane	+	water
CO_2	+	$4H_2$	$\rightarrow$	CH_4	+	$2H_2O$

Some methanogen species can survive extreme environments such as hot springs, hydrothermal vents, hot desert soil, and deep subterranean environments where there is a source of carbon. They are present in oil fields, coal beds, and in oil sands deposits. Jones et al. (2008) describe evidence that methanogens are responsible for the degradation of hydrocarbons in petroleum deposits, resulting in the formation of heavy oils.

Plants and Metals

Plants called *hyperaccumulators* absorb metals in soil through their root system and accumulate them in their leaves and upper organs (van der Ent et al. 2018). The ability of these plants to absorb metals can be performed in two processes. One is *phytomining* (a.k.a. *agromining*) in which hyperaccumulators are grown on a substrate having sufficient metal concentration and then harvested and processed to extract the metals. Processing involves burning the plants after harvest to create ash ("bio-ore") from which the metal is recovered by smelting or by application of reagents. Energy can be recovered from the burning. The other process is *phytoremediation* in which metals in contaminated soils are absorbed into plants that are then harvested and stored in secure sites. Figure 8.7 illustrates the two processes.

FIGURE 8.7

Phytomining and phytoremediation processes

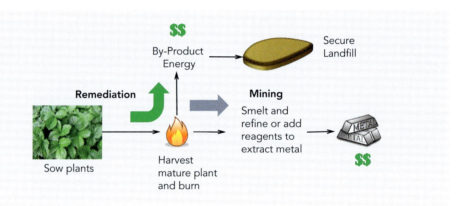

Several examples and applications of phytomining and phytoremediation have been described (Anderson et al. 2005; Sheoran et al. 2013; Hunt et al. 2014). Methods for commercial operations are described in van der Ent et al. (2018). Plant growth is a slow process, but the growth of hyperaccumulators can be accelerated by soil amendments, gene transfer from faster-growing species, or genetic engineering techniques.

For plants to accumulate metals, the metals must be in solution and available to the plant root system. However, metals such as gold, silver, and lead are relatively immobile and insoluble. Therefore, a process that puts these metals into solution (oxidizes them) must occur. Some plants, such as barley and maize, are cyanogenic; that is, they produce small quantities of cyanide that will oxidize metals. Bacteria in soils will excrete organic acids capable of oxidizing metals, possibly as a precursor to sequestration. These are particular cases. The underlying molecular mechanisms of metal accumulation in hyperaccumulators have evolved from basic mechanisms for metal accumulation in all plants, and genetic sequencing may provide a better understanding of these basic mechanisms (Merlot et al. 2018).

Metal accumulation by plants is believed to be a defense against herbivores and pathogens such as mildew and fungi. Indeed, some herbivores avoid plants containing metal, and larval-stage insects that feed on plants containing metal do not mature (Rascio and Navari-Izzo 2011). But some herbivores have adapted to this by developing the ability to distinguish metal concentrations and eating only those parts of the plant with low metal concentrations. Further investigation is needed to find ways to avoid herbivore damage to hyperaccumulator crops.

ADVANCED MACHINES AND SYSTEMS

Advanced technologies developed in other contexts or industries have been put to use in the mining industry with interesting results. Some are described next. Often these developments were relatively easy to implement, but equally often some imagination and unconventional thinking were needed. It is worthwhile to monitor developments in other fields—the problems or goals in these fields can be amazingly similar to those in the mineral resources industry.

Dancers, Drones, Data, and Digital Duals

Several companies and research groups have developed some interesting robot systems and applications. *Atlas* and *Spot* are two robots developed by Boston Dynamics (2023a, 2023b, 2023c). Atlas is an extreme athlete and a very good dancer to boot. Spot is a robotic dog used or being considered for use at some mines for safety inspection and remote measurement. Both robots have the ability to traverse rough terrain, climb stairs, and maneuver around or over obstacles. The mineral resources industry is just beginning to explore applications of this kind of technology.

Drones have been used for airborne geophysical exploration for some time. They are also used for inspection above ground (e.g., pit slopes) and below ground (e.g., stope surveys). Images from cameras on the drones are used to produce fully three-dimensional models of a mine.

Sensors on drones, such as those mentioned in the preceding subsection "Path 2. Capital Productivity," and described in Robben and Wotruba (2019), can be used to detect changes in rock properties or mineralization. More generally, data collection on a massive scale is possible with different kinds of sensors mounted on highly mobile robotic devices. The data can be used to construct a dual or *digital twin* of the mines and machines in a mining operation. These models are dynamic because they can be updated in real time to capture changes. Patterns of twin behavior can be detected and used to predict future behavior, useful in preventive maintenance and planning. Singh et al. (2022) review applications of digital twins to different industries including mining.

Robot Swarms

Can the biotechnologies described in this chapter be used to accumulate metals from a variety of sources such as metal scrap, mineralized zones in rock, or metal-contaminated water? Perhaps, but these processes are slow—very slow. We could be waiting a long time for a quantity of metal if such methods were used.

Nature combats the slowness by using numbers—very large numbers. Prime examples of this are insect colonies in which various tasks, such as foraging or defense, are undertaken by thousands of insects, each responding to environmental cues or to chemical signals from other insects. There is no "leader" insect or evidence of any kind of organization, yet the tasks are accomplished. It may seem inefficient in terms of modern project management methods, but it has the advantage of being able to adapt to any change and is therefore very robust.

Large numbers of insects are sometimes called *swarms*. Indeed, insect behavior when defending the colony does appear to be an overwhelming response to the threat in terms of numbers. Several researchers have worked on emulating swarm-like behavior by constructing large numbers of devices that are individually only capable of very simple operations but can collectively carry out a task by means of basic interactions. This collective behavior of a swarm is called *emergence*; that is, the behavior emerges from simple interactions; it is not programmed or established a priori.

A group at Harvard University recently developed a swarm of 1,024 small, inexpensive devices called the "Kilobot" (Rubenstein et al. 2014). The swarm is shown at the top of Figure 8.8. Each device has only a small circuit board, a pair of vibrating motors to power three rigid legs used for locomotion, and an infrared transmitter and receiver to allow messaging and localization. Three primitive behaviors are programmed into each device: (1) the ability to follow the edge of a group, (2) tracking the distance from an origin, and (3) maintaining a sense of relative location between neighbors.

The three behaviors are linked by an algorithm that uses the simple capabilities of each device to allow the swarm to self-assemble into a particular shape. Initially, the devices are placed in an arbitrarily shaped group and provided with an image

FIGURE 8.8

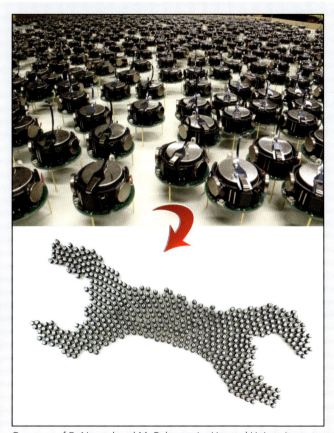

A swarm of Kilobots that self-assembled into the shape of a wrench

Courtesy of R. Nagpal and M. Rubenstein, Harvard University

of the desired shape. Four "seed robots" are placed next to the group to mark the place and orientation of the shape. The self-assembly algorithm then proceeds to move the robots into the shape. At the bottom of Figure 8.8 is an example of self-assembly into a wrench shape that the swarm completed in about 12 hours.

It should be possible to program these devices or similar ones so they could self-assemble at desired locations in an area and deploy some kind of reagent, consortium of microbes, or other biotechnology at the location. Whether this can be applied to the highly unstructured environments where minerals of interest occur remains a subject for investigation.

For more information on robot swarms, including a video showing Kilobot and six other robot swarms, and examples of potential applications of swarms to mining, see Tech Planet (n.d.), OffWorld (2023), Dyson (2021), and Anglo American (2023).

IN EXTREMIS: A VIRTUAL METAL SUPPLY COMPANY

The metal supply system shown in Figure 7.2 evolved during the 20th century as a result of the enormous metal demands of a rapidly growing global economy and the availability of large ore deposits. It is an efficient and effective organizational model for such deposits. Underlying the model at all levels is the concept of economies of scale; for example, large mining operations use large machines to spread fixed costs over large output, thus reducing unit costs. However, this is too inflexible to adapt to changes or to the variability of alternative sources of metals. Moreover, the system has managed to disconnect society from the metals and materials it uses. If we are to supply the metals needed for economic growth and for the energy transition, a much more flexible organizational model is needed. More importantly, the model must allow opportunities for society to become more engaged in the supply of metals and materials.

An alternative that would be possible using some of the technologies just described is more of an open organization where a variety of metal sources and extraction processes are available. All mineral and metal production processes and associated services would be performed by suppliers or dedicated service providers resulting in a virtual metal or materials supply system. Many opportunities for different companies, individuals, and communities to become involved in metal and materials supply would arise. A variant of this is illustrated in Figure 8.9. The main activity of the organization would be to manage the linkages between sources, processes, suppliers, and customers. Or perhaps it could manage itself, an autonomous metal supply system that knows when and where metals are needed.

Now it's your turn to think of what the future of mining could be. Think like there is no box.

FIGURE 8.9

Is this the future of mining?

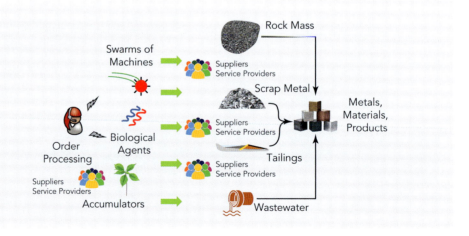

REFERENCES

Anderson, C., Moreno, F., and Meech, J. 2005. A field demonstration of gold phytoextraction technology. *Minerals Engineering* 18:385–392.

Anglo American. 2023. Picture this: The modern mine. www.angloamerican.com /futuresmart/stories/our-industry/technology/picture-this-the-modern-mine. Accessed November 2023.

Australian Academy of Science. 2012. *Searching the Deep Earth: A Vision for Exploration Geoscience in Australia*. Prepared by the UNCOVER group. Canberra, ACT: Australia Academy of Science. www.science.org.au. Accessed November 2023.

Baffes, J., and Nagle, P. 2022. Commodity demand: Drivers, outlook, and implications. In *Commodity Markets: Evolution, Challenges, and Policies*. Edited by J. Baffes and P. Nagle. Washington, DC: World Bank Group. pp. 121–183. doi:10.1596/978-1-4648-1911-7.

Bartos, P.J. 2007. Is mining a high-tech industry? Investigations into innovation and productivity advance. *Resources Policy* 32:149–158.

Boston Dynamics. 2023a. Atlas and beyond: The world's most dynamic robots. https://bostondynamics.com/atlas/. Accessed November 2023.

Boston Dynamics. 2023b. Case Study: Kidd Creek Mine. https://bostondynamics.com /case-studies/kidd-creek-mine/. Accessed November 2023.

Boston Dynamics. 2023c. Spot—The agile mobile robot. https://bostondynamics.com /products/spot/. Accessed November 2023.

Brune, K.D., and Bayer, T.S. 2012. Engineering microbial consortia to enhance biomining and bioremediation. *Frontiers in Microbiology* 3:1–6.

Dunbar, W.S. 2017. Biotechnology and the mine of tomorrow. *Trends in Biotechnology* 35(1):79–89.

Dunbar, W.S., and Klein, B. 2002. Mining, mineral processing and mini-machines. *CIM Bulletin* 1095:72–76.

Dyson, N. 2021. Enter the swarm. *Australia's Mining Monthly*. May 21. www.mining monthly.com/technology-innovation/news/1410634/enter-the-swarm. Accessed November 2023.

Earth Microbiome Project. 2023. Home page. www.earthmicrobiome.org. Accessed August 2023.

Gilbert, J.A., Jansson, J.K., and Knight, R. 2014. The Earth Microbiome project: Successes and aspirations. *BMC Biology* 12:69.

Hazen, R.M., and Ferry, J.M. 2010. Mineral evolution: Mineralogy in the fourth dimension. *Elements* 6:9–12.

Hunt, A.J., Anderson, C.W.N., Bruce, N., et al. 2014. Phytoextraction as a tool for green chemistry. *Green Processing and Synthesis* 3(1):3–22.

IEA (International Energy Agency). 2021. *Net Zero by 2050: A Roadmap for the Global Energy Sector*, 4th revision. Paris: IEA. https://iea.blob.core.windows .net/assets/deebef5d-0c34-4539-9d0c-10b13d840027/NetZeroby2050 -ARoadmapfortheGlobalEnergySector_CORR.pdf. Accessed November 2023.

IEA (International Energy Agency). 2023. Critical minerals data explorer. www.iea.org /data-and-statistics/data-tools/critical-minerals-data-explorer. Accessed October 2023.

ICSG (International Copper Study Group). n.d. Home page. https://icsg.org. Accessed October 2023.

IPCC (Intergovernmental Panel on Climate Change). 2018. *Global Warming of 1.5°C: An IPCC Special Report on the Impacts of Global Warming of 1.5°C Above Pre-Industrial Levels and Related Global Greenhouse Gas Emission Pathways, in the Context of Strengthening the Global Response to the Threat of Climate Change, Sustainable Development, and Efforts to Eradicate Poverty*. Cambridge, UK: Cambridge University Press. doi:10.1017/9781009157940.

Johnson, D.B. 2015. Biomining goes underground. *Nature Geoscience* 8:165–166.

Jones, D.M., Head, I.M., Gray, N.D., et al. 2008. Crude-oil biodegradation via methano-genesis in subsurface petroleum reservoirs. *Nature* 451:176–181.

Kaksonen, A.H., Deng, X., Bohu, T., et al. 2020. Prospective directions for biohydrometal-lurgy. *Hydrometallurgy* 195:105376.

Levett, A., Gleeson, S.A., and Kallmeyer, J. 2021. From exploration to remediation: A microbial perspective for innovation in mining. *Earth Science Reviews* 216:103563.

MMAP (Mining Microbiome Analytics Platform). 2023. Creating break-through biomin-ing solutions for natural resource extraction and green site remediation. https://www .digitalsupercluster.ca/projects/mining-microbiome-analytics-platform/. Accessed November 2023.

Merlot, S., de la Torre, V.S.G., and Hanikenne, M. 2018. Physiology and molecular biol-ogy of trace element hyperaccumulation. In *Agromining: Farming for Metals: Extracting Unconventional Resources Using Plants*, 2nd ed. Edited by A. van der Ent, A.J.M. Baker, G. Echevarria, et al. Cham, Switzerland: Springer Nature. p. 107.

NHGRI (National Human Genome Research Institute). n.d. The human genome project. Washington, DC: National Institutes of Health/NHGRI. www.genome.gov/human -genome-project. Accessed October 2023.

OffWorld. 2023. Extracting critical materials on earth and in space using swarms of indus-trial robots. www.offworld.ai/. Accessed November 2023.

Rascio, N., and Navari-Izzo, F. 2011. Heavy metal hyperaccumulating plants: How and why do they do it? And what makes them so interesting? *Plant Science* 180:169–181.

Reith, F., Rogers, S.L., McPhail, D.C., et al. 2006. Biomineralization of gold: Biofilms on bacterioform gold. *Science* 313:233–236.

Richards, J.P. 2013. Giant ore deposits formed by optimal alignments and combinations of geological processes. *Nature Geoscience* 6:911–916.

Robben, C., and Wotruba, H. 2019. Sensor-based ore sorting technology in mining—Past, present and future. *Minerals* 9(9):523.

Rubenstein, M., Cornejo, A., and Nagpal, R. 2014. Programmable self-assembly in a thousand-robot swarm. *Science* 345:795–799.

Schodde, R. 2014. Key issues affecting the time delay between discovery and develop-ment—Is it getting harder and longer? Slide presentation. https://minexconsulting .com/key-issues-affecting-the-time-delay-between-discovery-and-development-is-it -getting-harder-and-longer/. Accessed November 2023.

Sheoran, V., Sheora, A.S., and Poonia, P. 2013. Phytomining of gold: A review. *Journal of Geochemical Exploration* 128:42–50.

Singh, M., Srivastava, R., Fuenmayor, E., et al. 2022. Applications of digital twin across industries: A review. *Applied Sciences* 12:5727. doi:10.3390/app12115727.

Stewart, A.D., Anand, R.R., and Balkau, J. 2012. Source of anomalous gold concentrations in termite nests, Moolart Well, Western Australia: Implications for exploration. *Geochemistry: Exploration, Environment, Analysis* 12:327–337.

Tech Planet. n.d. 7 Incredible Swarm Robots. YouTube video, 5:33. www.youtube.com /watch?app=desktop&v=TYaquGrGhfk&feature=youtu.be. Accessed November 2023.

Teck. 2021. Spotlight on vehicle electrification at Teck. February 11. www.teck.com/news /stories/2021/spotlight-on-vehicle-electrification-at-teck. Accessed October 2023.

Thompson, L.R., Sanders, J.G., McDonald, D., et al. 2017. A communal catalogue reveals Earth's multiscale microbial diversity. *Nature* 551(7681):457–463.

UNFCC (United Nations Framework Convention on Climate Change). 2016. The Paris agreement. www.un.org/en/climatechange/paris-agreement. Accessed September 2023.

van der Ent, A., Baker, A.J.M., Echevarria, G., et al., eds. 2018. *Agromining: Farming for Metals: Extracting Unconventional Resources Using Plants*, 2nd ed. Cham, Switzerland: Springer Nature.

Volvo. 2023. Volvo trucks and Boliden collaborate on deployment of underground electric trucks for mining. News release March 21. www.volvotrucks.com/en-en/news-stories /press-releases/2023/mar/volvo-and-boliden-collaborate-on-electric-underground -trucks.html. Accessed October 2023.

All the Chemistry You Need to Know

Some aspects of how minerals are converted to metals, how metals are refined, and how metals behave in the environment are more easily explained, often with a few pictures, if some simple aspects of chemistry are understood. You may have seen a lot of what follows in high school.

ATOMS AND IONS

Figure A.1 shows an atom consisting of a central nucleus surrounded by electrons that are in orbits about the nucleus. The nucleus contains positively charged protons and neutrons that have no charge. There are as many electrons as there are protons, giving the atom zero net charge. The atom shown in Figure A.1 has 11 electrons and therefore has 11 protons in its nucleus. This is the atomic configuration of the metal sodium (chemical symbol Na). For comparison, copper has 29 electrons and gold has 79 electrons.

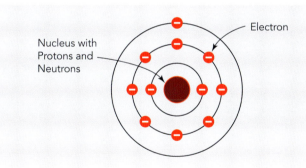

FIGURE A.1

Sodium atom with 11 electrons

Other than the fact that it has a number of positively charged particles (protons) equal to the number of negatively charged particles (electrons), the nucleus is not of interest in mineral processing and refining. The electrons are of interest because it is the electrons that are involved in the chemical reactions used to refine metals.

Some atoms or molecules give up electrons and become positively charged because when electrons leave, there are more protons than electrons. Others take on electrons and become negatively charged. When an atom gives up or takes on electrons, it is called an *ion*.

Sodium easily gives up the single electron in its outer orbit, resulting in a positive ion denoted Na$^+$ and illustrated in Figure A.2. The chemical equation for the loss of the electron is

$$Na \rightarrow Na^+ + e$$

Sodium ion Na$^+$ with 10 electrons, one less than the atom Na

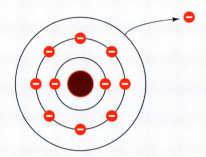

Another atom is chlorine (symbol Cl), which has 17 electrons, 7 of which are in its outer orbit. The outer orbit of chlorine needs 8 electrons and so it will pick up the extra electron from wherever it can, leading to a negative ion Cl$^-$, as illustrated in Figure A.3. The chemical equation for the gain of the electron is

$$Cl + e \rightarrow Cl^-$$

FIGURE A.3

Chlorine ion Cl$^-$ with 18 electrons, one more than the atom Cl

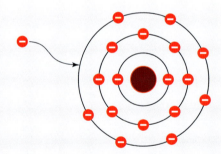

Ions exist only in a liquid medium such as water; they are said to be in solution. Anyone who has used salt, a compound with the chemical formula NaCl, has made the sodium and chlorine ions. As soon as the compound NaCl (sodium chloride) encounters a fluid on food, it breaks up (dissolves) into the two ions Na$^+$ and Cl$^-$ (Figure A.4). The chemical equation for this is

$$NaCl(s) \rightarrow Na^+(aq) + Cl^-(aq)$$

where the "s" denotes solid and the "aq" denotes in aqueous or water solution—as an ion. Salt is highly soluble in water.

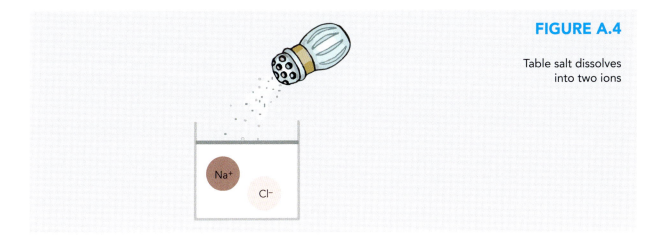

Some atoms can give up or take on more than one electron. For example, when magnesium (symbol Mg) gives up the two electrons in its outer orbit, the ion Mg^{++} (or Mg^{2+}) is produced. Oxygen will take two electrons into its outer electron orbit to become O^{2-}.

OXIDATION AND REDUCTION

When an atom or molecule gives up an electron, it is "oxidized," and when it takes on an electron, it is "reduced." The reason for the term *oxidation* is that the oxygen atom has an affinity for electrons—it will take on two electrons into its outer electron orbit, even from an atom that does not easily gives up electrons. The term *reduced* means that the number of positive charges (which could be zero) has been decreased.

Other chemicals that do not contain oxygen but are strong oxidizers are used in mineral processing and refining. One example is the *ferric* ion Fe^{3+}, which will take an electron from another ion or molecule and become the *ferrous* ion Fe^{2+}. The ion or molecule is oxidized and the ferric ion is reduced because its number of positive charges is reduced from three to two.

Some metals oxidize more easily than others. Table A.1 shows some metals arranged in order of their increasing ability to oxidize.

Increasing ease of oxidation (forming the ion shown) →

Gold	Platinum	Silver	Copper	Lead	Nickel	Iron	Zinc
Au^+	Pt^{2+}	Ag^+	Cu^{2+}	Pb^{2+}	Ni^{2+}	Fe^{2+}	Zn^{2+}

TABLE A.1

Metals and
oxidation

Gold, platinum, and silver are sometimes called *noble metals* because they do not oxidize easily nor do they engage in chemical reactions. Copper, lead, and zinc do oxidize easily. These differences in the ability to oxidize have implications for methods of processing and refining these metals.

WHY DON'T METAL SULFIDES DISSOLVE IN WATER?

The answer is they will, but very slowly and depending on conditions. It would be great if metal sulfides, such as chalcopyrite ($CuFeS_2$), sphalerite (ZnS), and galena (PbS), dissolved in water as fast as salt does because then the metal ions would be in solution and pure metals could be obtained by electrolysis of the solution. However, given the high availability of water in the earth, if metal sulfides were as soluble as salt, a lot of metal ions would be in groundwater, oceans, or lakes—which is not good for life as we know it.

Salt is an ionic compound in which the bond between the sodium ion Na^+ and the chlorine ion Cl^- is formed by the electrostatic attraction between oppositely charged particles. The chlorine atom has taken the electron the sodium atom gave up. The reason salt dissolves very easily in water is because the water gets into the salt crystals and acts as a kind of barrier between the charged particles, preventing one from "seeing" the other. Metal sulfides are not ionic compounds. In fact, the atoms in metal sulfides are sharing electrons and are therefore more tightly bound to each other. These bonds are not easily broken.

Under acidic conditions, high temperatures, or in the presence of chemicals that are oxidizers, metal sulfides will dissolve and lead to acid rock drainage and metal contamination. This can occur naturally or it may be caused by the exposure of sulfide minerals during mining activities. Certain types of bacteria accelerate the breakdown of sulfide minerals as a source of energy for growth. This also puts metals into solution.

THE MAIN MESSAGE

The goal of processing and refining metals is to get the metals into solution as positive ions.

Once they are in solution, electricity, which is a flow of electrons, can be used to add electrons to the metal ions to make them into pure atoms, a process called *electrolysis*. This is one way of refining a metal, and it is accomplished by "plating" the metal onto a solid surface.

That's all the chemistry you need to know to understand mineral processing and refining. Really!

KITCHEN CHEMISTRY

Two relatively simple experiments can be done in your own kitchen to demonstrate a few concepts regarding copper plating.

Experiment 1

The experiment shown in Figure A.5 illustrates the concept of ions in solution and plating a metal. This is a small-scale version of a technique used in industry to obtain copper from solutions of copper sulfate ($CuSO_4$) derived from leaching low-grade copper ore.

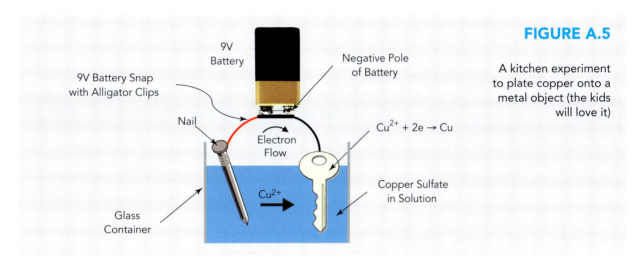

9V Battery

9V Battery Snap with Alligator Clips

Negative Pole of Battery

Nail

Electron Flow

$Cu^{2+} + 2e \rightarrow Cu$

Cu^{2+}

Copper Sulfate in Solution

Glass Container

FIGURE A.5

A kitchen experiment to plate copper onto a metal object (the kids will love it)

At the following link is a description (including a video) of how to perform the preceding experiment. The problem you might encounter is obtaining the copper sulfate. It is used in root killer, but some jurisdictions have banned its sale. All the other required materials can be easily obtained.

www.homesciencetools.com/article/electroplating-science-project/

If you are able to get the copper sulfate and can do the experiment, this could be the result:

Think of the many things in your possession that need plating like this!

Experiment 2

This experiment, described at the following link, uses vinegar (an acid) to strip part of the copper plating off pennies, which puts copper ions in solution. The copper ions will then plate onto nails placed into the solution.

http://homechemistry.blogspot.ca/2008/01
/penny-chemistry-verdigris-and-copper.html

The problem might be finding the pennies!

ACID AND ALKALI SOLUTIONS

Two other terms to know are *acid* and *alkali*. The acidity or alkalinity of a solution is measured by the pH scale, which is a measure of the concentration of the hydrogen ion H^+, a constituent of all acids. The pH scale ranges from 0 to 14. A pH of 7 is *neutral*, a pH less than 7 is acidic, and a pH greater than 7 is alkali. The pH values of some common liquids are shown on the scale in Table A.2.

TABLE A.2

The pH scale

pH	Example
0	Battery acid
1	Gastric acid (in stomach)
2	Lemon juice, vinegar
3	Fruit juice
4	Tomato juice
5	Black coffee
6	Urine, saliva
7	Pure water
8	Sea water
9	Baking soda solution
10	Great Salt Lake, Utah
11	Ammonia solution
12	Soapy water
13	Bleach, oven cleaner
14	Liquid drain cleaner

Grade Distributions and Grade–Tonnage Curves

DISTRIBUTION OF METAL GRADES

Figure B.1 shows three histograms of metal grades measured by means of assays of blasthole cuttings or channel samples along a vein.

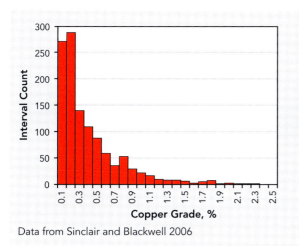

Number of data: 1,181
Mean grade: 0.36%
Median grade: 0.22%
Minimum grade: 0.002%
Maximum grade: 2.78%

FIGURE B.1a

Similkameen blasthole data (British Columbia, Canada)

Data from Sinclair and Blackwell 2006

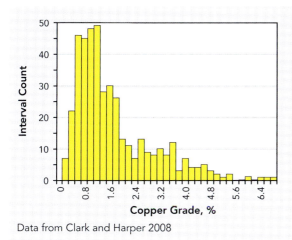

Number of data: 428
Mean grade: 1.63 g/t
Median grade: 1.20 g/t
Minimum grade: 0.07 g/t
Maximum grade: 6.89 g/t

FIGURE B.1b

Sunshine Mine gold vein (Idaho, United States)

Data from Clark and Harper 2008

FIGURE B.1c

Zinc vein in
Pulacayo mine
(Bolivia)

Number of data: 118
Mean grade: 15.61%
Median grade: 13.65%
Minimum grade: 3.7%
Maximum grade: 39.3%

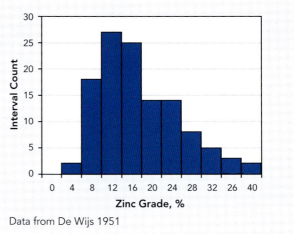

Data from De Wijs 1951

These histograms are constructed by dividing the range between the minimum and maximum grade into a number of intervals of equal size and then counting the number of grades that are within each interval. The height of each vertical bar in the histogram represents the number of grades that lie within the interval represented by the width of the bar. For example, Figure B.1a shows 273 values lying between 0% and 0.1% and 289 values lying between 0.1% and 0.2%.

The Similkameen blastholes were drilled on the top of a bench in an open pit mine in British Columbia (Canada). If the block of ore under the bench is assumed representative of the ore body, then an estimate of the percentage of ore below a particular grade g, $P(\leq g)$, may be made by adding up all the interval counts up to and including g and dividing the sum by the number of measurements. The result is shown in Figure B.2. The percentage of ore above grade g is given by $1 - P(\leq g)$, and this is also shown in Figure B.2. The graphs in Figure B.2 are known as *grade–tonnage curves*.

From the percentage above grade graph, it can be seen that the available reserves above a particular grade decrease rapidly as the grade approaches the average grade. If the cutoff grade is equal to the average grade 0.36%, shown by the vertical dashed line, the horizontal dashed line shows that only 35% of the reserves have a grade equal to or greater than 0.36%.

The metal grades in the three histograms each exhibit a preponderance of lower grades. A simple interpretation is that low grades are more probable than high grades. One can (and one will) speculate why this is the case. The minerals were precipitated from a hot flowing fluid. One possibility is that as mineral precipitation begins at one point in the fluid, the physical and chemical conditions necessary

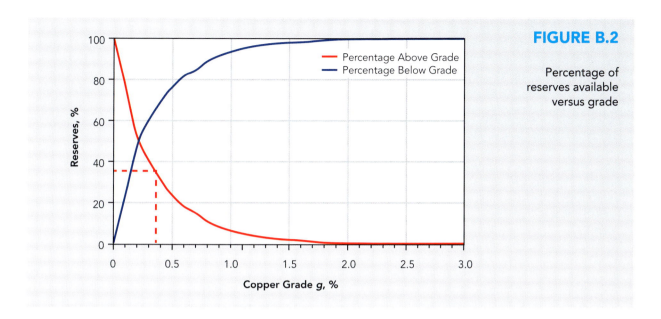

FIGURE B.2

Percentage of reserves available versus grade

for precipitation are interrupted by the flow and mixing of the fluids. This likely happens more often than not and so high grades are rare.

REFERENCES

Clark, I., and Harper, W.V. 2008. *Practical Geostatistics 2000 Case Studies*. Alloa, Scotland: Geostokos (Ecosse) Limited.

De Wijs, H.J. 1951. Statistics of ore distribution, part I. *Geologie en Mijnbouw* 30:365–375.

Sinclair, A.J., and Blackwell, G.H. 2006. *Applied Mineral Inventory Estimation*. Cambridge, UK: Cambridge University Press.

Index

Note: *f.* indicates figure; *t.* indicates table